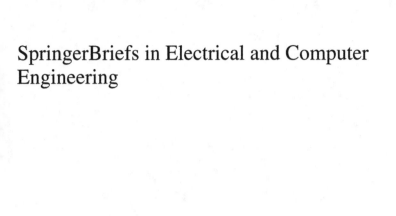

SpringerBriefs in Electrical and Computer
Engineering

More information about this series at http://www.springer.com/series/10059

Changyan Yi • Jun Cai

Market-Driven Spectrum
Sharing in Cognitive Radio

 Springer

Changyan Yi
Department of Electrical
  and Computer Engineering
University of Manitoba
Winnipeg, MB, Canada

Jun Cai
Department of Electrical
  and Computer Engineering
University of Manitoba
Winnipeg, MB, Canada

ISSN 2191-8112   ISSN 2191-8120   (electronic)
SpringerBriefs in Electrical and Computer Engineering
ISBN 978-3-319-29690-6   ISBN 978-3-319-29691-3   (eBook)
DOI 10.1007/978-3-319-29691-3

Library of Congress Control Number: 2016931595

Printed on acid-free paper

This Springer imprint is published by SpringerNature
The registered company is Springer International Publishing AG Switzerland

Recommended by Sherman Shen

Preface

Radio spectrum has become a scarce and precious resource in wireless communications due to explosive growth of demand from newly developed wireless devices, applications, and services. Meanwhile, the existing rigid spectrum regulatory policy, which is based on static allocation with exclusive spectrum usage, has led to significant spectrum under-utilization. In order to alleviate the burden of spectrum shortage, dynamic spectrum sharing based on cognitive radio (CR) has been introduced recently to redistribute spectrum in a more intelligent and flexible way. Among various methods for implementing dynamic spectrum sharing, market-driven sharing schemes, especially the algorithmic mechanism design approach, has been widely accepted as the prospective solution thanks to its advantages in offering allocation efficiency, fairness, and economic incentives. The objective of this brief is to present basic ideas and unique challenges involved in adopting algorithmic mechanism for the purpose of dynamic spectrum sharing and examine recent advances in this area. The network architecture of CR and the characteristics of dynamic spectrum sharing are first explored, followed by the review of fundamentals of mechanism design theory and its potential deploying scenarios. Then, the readers are exposed to the latest mechanism designs for dynamic spectrum sharing in CR networks by analytically devising three featured mechanisms, i.e., recall-based spectrum auction mechanism, two-stage spectrum sharing mechanism, and online spectrum allocation mechanism. Numerical results are provided to demonstrate the feasibility and significant performance enhancements in spectrum utilization efficiency by adopting these mechanism design approaches.

This brief is a concise and approachable material in the design of dynamic spectrum sharing mechanisms, and it is hopefully found to be helpful for researchers, professionals, and advanced-level students in the fields of wireless communications and networking.

We would like to specially thank the series editor, Prof. Xuemin (Sherman) Shen from University of Waterloo, for encouraging us to prepare this brief. Besides, our sincere thanks goes to all the colleagues in Communication and Network Engineering Research Group at the University of Manitoba, for their kind suggestions and continuous help. Finally, we would also like to thank Natural

Sciences and Engineering Research Council of Canada (NSERC) Discovery Grant for the funding support.

Winnipeg, MB, Canada   Changyan Yi
February 2016   Jun Cai

Contents

Acronyms

BIP	Binary integer programming
CR	Cognitive radio
DSA	Dynamic spectrum access
FCC	Federal Communications Commission
FCFS	First-come-first-serve
GPS	Global positioning system
LOS	Lehmann–Oćallaghan–Shoham
LO	Local oscillator
MAC	Medium access control
MKP	Multiple knapsack problem
NE	Nash equilibrium
NP	Non-deterministic polynomial time
OSI	Open systems interconnection
PBS	Primary base station
PO	Primary spectrum owner
PU	Primary user
QoS	Quality-of-service
RSSA	Recall-based single-winner spectrum auction
RMSA	Recall-based multiple-winner spectrum auction
SBS	Secondary base station
SDR	Software defined radio
SPSB	Second-price sealed-bid
SPTF	Spectrum Policy Task Force
SNR	Signal-to-noise ratio
SU	Secondary user
TAGS	Two-stage resource allocation mechanism with combinatorial auction and Stackelberg game in spectrum sharing
VCG	Vickrey–Clarke–Groves
WDP	Winner determination problem

Chapter 1
Introduction

1.1 Background and Aim of the Brief

With the development of technologies in various fields of electrical and computer engineering, wireless communication has become increasingly popular and even indispensable in our daily life. Electronic devices or equipments, such as mobile phones, tablets, global positioning system (GPS), cordless computer peripherals, remote controllers, and satellite televisions, are all based on wireless communication technology. As a consequence, wireless technology has experienced rapid evolution in the past decade, and has been attracting more and more research, application, and business interests from both academia and industry.

Generally, wireless communication relies on the use of radio, which refers to the transmission of electromagnetic energy through space. Radio is able to carry information with proper modulation of radiated waves in terms of their amplitudes, frequencies, and phases. These radio waves can be transmitted and received through antennas, which convert the radio waves into electric currents, and vice versa. Radio spectrum, as the media of communication, commonly occupies the frequency range from 3 kHz to 300 GHz. Different frequency bands possess heterogeneous propagation characteristics, and fit different applications of wireless services. For instance, low-frequency radio bands are more suitable for long-range communications, while high-frequency bands are more suitable for short range but high-speed wireless transmissions.

Considering that radio spectrum is an essential and necessary resource for wireless communications, how to efficiently allocate radio spectrum becomes an everlasting problem since the born of wireless technology. Signal interference was once considered as the main factor in spectrum allocation. Specifically, interference can occur when multiple radios transmit simultaneously over the same frequency. Hence, traditional spectrum management statically assigns exclusive spectrum bands to different wireless users to avoid potential interference. Since 1930s,

© The Author(s) 2016
C. Yi, J. Cai, *Market-Driven Spectrum Sharing in Cognitive Radio*,
SpringerBriefs in Electrical and Computer Engineering,
DOI 10.1007/978-3-319-29691-3_1

Table 1.1 Licensed spectrum allocation in the USA [2]

Wireless services	Frequency bands
AM radio	535–1605 kHz
FM radio	88–108 MHz
Broadcast TV (Channel 2–6)	54–88 MHz
Broadcast TV (Channel 7–13)	174–216 MHz
Broadcast TV (UHF)	470–806 MHz
Broadband wireless	746–764 MHz, 776–794 MHz
3G wireless	1.7–1.85 GHz, 2.5–2.69 GHz
1G and 2G digital cellular	806–902 MHz
Personal communications service	1.85–1.99 GHz
Wireless communications service	2.305–2.32 GHz, 2.345–2.36 GHz
Satellite digital radio	2.32–2.325 GHz
Multichannel multipoint distribution service (MMDS)	2.15–2.68 GHz
Digital broadcast satellite (Satellite TV)	12.2–12.7 GHz
Local multipoint distribution service (LMDS)	27.5–29.5 GHz, 31–31.3 GHz
Fixed wireless services	38.6–40 GHz

spectrum has been assigned through administrative licensing by governments. For example, Federal Communications Commission (FCC) in USA adopts the command-and-control management approach, and acts as a centralized authority to determine the spectrum allocation and usage by granting licenses to authorized parties [1]. Such allocation pattern is normally static in both temporal and spatial dimensions. In other words, spectrum licenses are valid for ages (usually decades) and for large geographical areas (country wide). As an example, the licensed spectrum allocation for different major wireless services in the USA is demonstrated in Table 1.1.

Though the command-and-control based management framework ensures exclusive spectrum usages, and thus guarantees interference free communications, it has been argued as an artifact of outdated technologies due to its simplicity and inflexibility. In addition, as claimed in the report from the spectrum Policy Task Force (SPTF) of the FCC in 2002 [3], the licensed spectrum bands are only utilized 15–85 % with a high variance in time, which implies a significant spectrum under-utilization. Moreover, with the explosive growth of demand for radio resource from newly developed wireless equipment and applications, the existing spectrum regulatory policy has imposed significant restrictions on spectrum utilization efficiency which leads to a serious issue, called *spectrum scarcity*.

In order to exploit under-utilized spectrum to meet the demand of future wireless applications/services, some recommendations in changing the spectrum management policy have been proposed [4]. These suggestions include: (1) increasing the flexibility of spectrum usage; (2) considering all dimensions and related issues of spectrum allocation into the policy; and (3) supporting and encouraging efficient use of spectrum. These recommendations should be understood and implemented from both technical and economic aspects. From the view of wireless techniques, spectrum management has to ensure low interference and high utilization of radio frequency bands. From the economic perspective, an incentive mechanism should be integrated in spectrum management so as to provide extra revenue and economic satisfaction for spectrum licensees.

During the past two decades, different spectrum management models have been introduced and extensively studied [5]. The common objective is to improve the flexibility of spectrum usage and efficiency of spectrum utilization. To achieve this goal, the wireless transceivers are expected to be more intelligent to access the radio spectrum, which motivates the development of a new wireless technology, called *cognitive radio* (CR). CR has been considered as a prospective dynamic spectrum access (DSA) solution to allow unlicensed wireless users to opportunistically access the licensed spectrum on the premise that the services of authorized users are not degraded because of interference [6, 7].

The concept of CR was first presented by Joseph Mitola III in 1999 [8]. It is formally defined in [9] as follows: *CR is an intelligent wireless communication system that is aware of its surrounding environment and uses the methodology of understanding-by-building to learn from the environment and adapt its internal states to statistical variations by adjusting the transmission parameters (e.g. frequency band, modulation mode, and transmit power) in real-time.* The main functions of CR include spectrum sensing, spectrum decision, spectrum sharing, and spectrum mobility. Through spectrum sensing, CR users (also called, secondary users) detect the information of licensed spectrum (e.g., the gain and the activities of users who own the spectrum). Then, sensing information is collected and used for making decisions on spectrum access. If the radio environment changes, CR users could change the frequency of operation by the function of spectrum mobility.

However, CR based on spectrum sensing [10, 11] has two inherent limitations: (1) Licensed users/primary users (PUs) are presumed to be unconscious of secondary users' (SUs') sensing activities so that PUs have no countermeasures even their interests are harmed. For example, SUs may violate PUs' interference tolerances due to misdetection. Since perfect sensing is impossible in practice, mitigating interference and impairments to PUs is always challenging; (2) The development of spectrum sharing scheme for CR requires the cooperation of PUs since most of the spectrum bands have already been allocated/sold to these licensed users. Intuitively, if adopting CR can only increase the spectrum utilization and flexibility (which are solely beneficial for SUs), self-interested PUs would not support and may even hinder the implementation of CR because their spectrum usages have already been guaranteed through pre-paid licenses and they may feel unfair against unpaid SUs.

For the reasons stated above, CR based on spectrum marketing [12] has attracted more and more attention recently. Compared to the sensing-based CR, PUs could take initiative in spectrum marketing by deciding the quantity of spectrum to be leased so that their utilities can be maximized within their interference tolerance. Moreover, instead of free sharing based on sensing, PUs could charge SUs for dynamically using licensed spectrum for their license costs and potential performance degradation. It is obvious that market-driven/pricing-based CR can not only enhance the spectrum efficiency, but also provide economic incentives for PUs to participate in DSA.

This brief focuses on the current research on mechanism design for market-driven dynamic spectrum sharing in CR networks. Along with a review of CR architectures and characteristics, this brief explains the motivations, significance, and unique challenges of implementing algorithmic mechanism design for encouraging the participation of both primary spectrum owners and secondary spectrum users in dynamic spectrum sharing. With an emphasis on dealing with the uncertain spectrum availabilities in CR networks, this brief introduces some recent advances of spectrum sharing mechanisms.

In Chap. 3, a recall-based spectrum auction mechanism is presented, where SUs can declare heterogeneous quality-of-service (QoS) requirements and the primary base station (PBS) can recall channels after auction to deal with the potential sudden increase in its own PUs' demands. Beginning with the illustration of a recall-based single-winner spectrum auction (RSSA) mechanism, the model is further extended to allow multiple winners in order to improve the spectrum utilization, which results in a recall-based multiple-winner spectrum auction (RMSA) mechanism. A combinatorial auction model is then formulated, and Vickrey–Clarke–Groves (VCG) mechanism is applied in the payment design. Moreover, the RMSA mechanism focuses on a fair spectrum allocation among heterogeneous SUs and the increase of the PBS's auction revenue. Both theoretical and simulation results show that the recall-based spectrum auction mechanisms can improve the spectrum utilization with guarantees on SUs' heterogeneous QoS requirements.

In Chap. 4, a two-stage resource allocation mechanism with combinatorial Auction and Stackelberg Game in spectrum Sharing (TAGS) is analyzed, where the DSA among multiple heterogeneous primary spectrum owners (POs) and SUs in recall-based CR networks is investigated. In this framework, SUs can demand a different amount of spectrum for their transmissions, and each PO will provide a portion of radio resources for leasing, but has to guarantee its own PUs a certain degree of QoS. Furthermore, POs are allowed to have different spectrum trading areas as well as heterogeneous activities among POs' users. In the first stage, a spectrum allocation is decided by running a geographically restricted combinatorial auction without the consideration of spectrum recall. In the second stage, a Stackelberg game is formulated for all users to determine their best strategies with respect to the potential spectrum recall. Both theoretical and simulation results prove that TAGS provides a feasible solution for the problem under consideration and ensures the desired economic properties for all individuals.

In Chap. 5, an online spectrum auction mechanism for CR networks with uncertain activities of PUs is illustrated. In this framework, a single PBS, acted as the spectrum auctioneer, leases its under-utilized channels to SUs who request and access spectrum on the fly. Moreover, this mechanism focuses on a more practical situation that the auctioneer (or the PBS) has no prior knowledge of PUs' activities so that its channel states are uncertain. In order to balance the auction profits from granted SUs' spectrum requests and the potential penalties caused by incomplete services to PUs, the idea of virtual spectrum sellers is introduced and the spectrum allocation problem is formulated as an online double auction. Then, a novel online admission and pricing mechanism is studied with the consideration of spatial reusability. Theoretical analyses are provided to prove the desired economic properties in terms of budget-balance, individual rationality, and incentive compatibility. Simulation results demonstrate the superiority of the online mechanism in increasing the utility of the PBS, improving the spectrum utilization, and providing better satisfaction for SUs compared to counterparts.

Finally, in Chap. 6, the brief concludes with a discussion of potential research directions and interests, which will motivate further studies on mechanism design for more general wireless communications.

1.2 Overview of Dynamic Spectrum Sharing

The severe spectrum scarcity and the inefficiency of current spectrum allocation have stimulated a flurry of researches in engineering, economics, and regulation communities for developing better spectrum management schemes. In this section, we first provide an overview of CR networks, which have been regarded as the key enabling technology to mitigate the spectrum under-utilization through DSA. Then, the general framework of DSA will be presented with detailed descriptions of its procedures, functionalities, and research challenges. Finally, we will illustrate the motivations and necessities of considering market-driven dynamic spectrum sharing, and demonstrate its basic structure and related issues.

1.2.1 Architecture of Cognitive Radio Networks

CR, pioneered by Joseph Mitola III from software defined radio (SDR), was originally considered as a strengthened SDR with artificial intelligence [8]. With such concept, CR was imagined to be capable of sensing the radio environment and reacting accordingly. FCC endorsed the idea of CR shortly and provided a more explicit definition [3]: *CRs are radios which could opportunistically use licensed spectrum bands under the restriction of interference temperature of PUs*. From this definition, CR has two main characteristics [9]:

- *Cognitive capability*: CR users can identify portions of unused spectrum through real-time interaction with radio environment. CR enables the opportunistic usage of temporally unused spectrum (also referred to as *spectrum hole* or *white space*) among SUs without interfering licensed users. A simple illustration is shown in Fig. 1.1 [13].
- *Reconfigurability*: CR users could transmit and receive on various radio frequencies, and apply different access technologies through software reconfiguration.

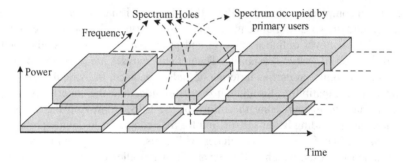

Fig. 1.1 Illustration of spectrum holes

The primary objectives of CR networks are: (1) improving spectrum utilization in a fair-minded way; and (2) providing highly reliable communications for all users in networks.

In order to develop efficient DSA schemes based on the CR technology, it is essential to understand the architecture of CR networks. Basically, the CR architecture can be described based on either the network classification or the framework of the standard open systems interconnection (OSI) model.

- *Network classification*: The components of a CR network can be classified into two categories [7], i.e., *primary networks* and *secondary networks*, as shown in Fig. 1.2. Primary networks are existing networks for PUs to operate on specific licensed spectrum bands. If primary networks have infrastructure, PBSs are commonly equipped to control the spectrum usage of PUs. Since PUs are always granted with higher spectrum access priorities, the performance of primary networks should not be affected by SUs' activities. On the other hand, secondary networks have no license for spectrum usage. Thus, additional functions are required for SUs to dynamically access the licensed spectrum bands. Secondary networks can also be equipped with secondary base stations (SBSs) to dynamically allocate the spectrum resources among SUs.
- *OSI framework*: The CR architecture can also be considered in a layered structure following the conventions of OSI model [14], as shown in Fig. 1.3. In physical and data link layers, *spectrum sensing* is essential in discovering spectrum holes as well as protecting PUs. *Resource allocation* and *CR medium access control (MAC) protocol* are designed to perform similar functions as in traditional wireless networks. However, in CR networks, these functions should be aware of and adapt to fluctuating spectrum availability, and thus require a collaboration between physical and data link layers. *Spectrum-aware opportunistic routing protocol* aims to manage the CR-based routing via cross-layer interactions of link and network layers so that the optimal route can be found by checking hop-by-hop spectrum availability. *CR transport protocol* can be seen as an improvement of traditional transport protocols with spectrum availability awareness. In the application layer, *spectrum trading* refers to the dynamic spectrum sharing between PUs and SUs in terms of market mechanisms. Besides, with the use

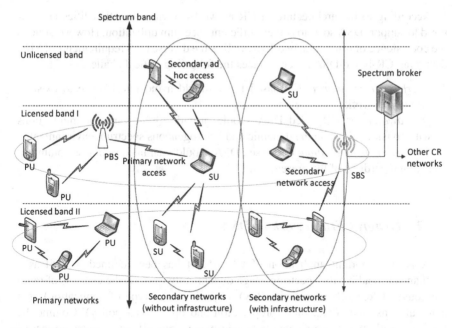

Fig. 1.2 CR architecture based on network classification

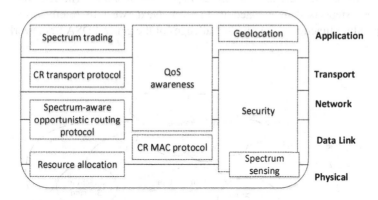

Fig. 1.3 CR architecture based on OSI model

of *geolocation* databases, a look-up table of PUs' spectrum usage (especially for highly predictable PUs' activities) can be built which provides easier ways to check the presence of PUs in different frequency bands. Finally, *QoS awareness* and *security* are also inherent CR functions that span over multiple layers, where the former provides solutions to heterogeneous QoS provisioning, and the latter protects both PUs and SUs from potential threats that can disrupt efficient operations of CR.

According to the architecture of CR networks, various functionalities are required to support DSA so as to achieve efficient spectrum utilization. However, due to the coexistence of primary networks as well as the diverse QoS requirements of SUs, designing CR-based DSA schemes faces the following critical challenges [7]:

1. *Interference avoidance*: CR-based DSA should avoid interference to existing primary networks.
2. *QoS awareness*: CR-based DSA should provide QoS-aware communications with considerations on the dynamic and heterogeneous spectrum environment.
3. *Seamless communication*: CR-based DSA should guarantee seamless communication regardless of the presence of PUs.

1.2.2   General Framework of DSA

DSA is an important application of CR, and it has been defined in [15] as a mechanism to adjust the spectrum usage dynamically towards the changes of radio environment (e.g., channel availability), objective (e.g., type of application), and external constraints (e.g., radio propagation and operational policy). Commonly, the CR-based DSA process consists of four major steps [7], i.e., *spectrum sensing*, *spectrum decision*, *spectrum sharing* and *spectrum mobility*. The relationship of these four steps is demonstrated in Fig. 1.4. Next, we will briefly discuss the corresponding functions along with challenges of the general DSA framework.

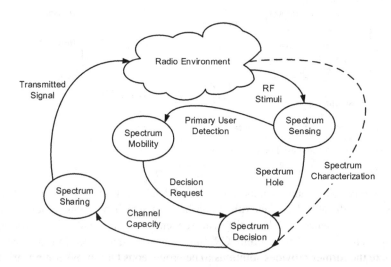

Fig. 1.4 CR cycle

1.2.2.1 Spectrum Sensing

Since SUs can only access unused spectrum of PUs, CR is required to monitor the licensed spectrum and detect the spectrum holes through periodical sensing. Spectrum sensing techniques can be further classified into three groups, i.e.,

- *Primary transmitter detection*: With local observations, SU can detect weak signals from primary transmitters. Currently, there are three major techniques for transmitter detection, i.e., matched filter detection, energy detection, and feature detection, each of which has its own specific application scenarios.
- *Primary receiver detection*: The most effective way to find spectrum holes is to detect PUs that are receiving data within the radio range of an SU. It can be achieved by sensing the local oscillator (LO) leakage signals of primary receivers. However, its implementation is considerably difficult because the LO leakage signals are extremely weak.
- *Interference-based sensing*: By measuring the noise levels at the receivers of PUs, SUs can control their spectrum access without violating the detected interference temperature limits.

However, some research challenges still remain unsolved in the development of spectrum sensing techniques:

- *Interference measurement*: The main objective of spectrum sensing is to obtain the accurate status of the licensed spectrum. The challenge lies in the interference measurement at the primary receiver caused by transmissions from SUs. For example, an SU may not know the primary receiver's precise location (which is required to compute the interference tolerance). In addition, if a primary receiver is a passive device, the transmitter may not be aware of the receiver [13].
- *Spectrum sensing in multiuser networks*: It is practical that multiple PUs and SUs may share the same radio spectrum in one network or in multiple networks with coexisting issues. Such multiuser environment makes it very difficult to sense spectrum holes and estimate interference.
- *Spectrum-efficient sensing*: Intuitively, a single-radio SU cannot perform both sensing and transmitting simultaneously in the same frequency band. Extending the sensing period can lead to a higher sensing accuracy, while result in a lower transmission throughput. Thus, balancing transmission and sensing durations is also a critical issue.

1.2.2.2 Spectrum Decision

The information collected from spectrum sensing is used to make spectrum allocation decisions. Such decisions are made by optimizing some desired performance metrics (e.g., total throughput of SUs) under the constraints of spectrum availabilities and interference limits. Spectrum decision process is challenged by the following open research issues:

- *Decision model*: The spectrum bands in CR networks are difficult to be characterized due to the complexities in analyzing parameters, such as the durations of spectrum holes and the signal-to-noise ratio (SNR) of the target frequency band. Besides, the heterogeneities in QoS provisioning of SUs should also be taken into account. Till now, designing an application- and spectrum-adaptive decision model is still difficult.
- *Competition/cooperation*: In multiuser CR networks, users can be cooperative or non-cooperative in accessing the spectrum. In a non-cooperative scenario, each user has its own objective, while in a cooperative scenario, all users can cooperate to achieve a common goal. Thus, it is necessary to exploit the preference of each user in order to make optimal decisions.
- *Spectrum decision over heterogeneous spectrum bands*: Since certain spectrum bands are assigned for different purposes, a CR network has to support spectrum decision operations on heterogeneous spectrum bands.

1.2.2.3   Spectrum Sharing

After a decision has been made, spectrum holes may be accessed by multiple SUs. In order to avoid possible collisions with licensed users and other unlicensed users, a cognitive MAC protocol should be carefully designed so as to efficiently share the spectrum resource. The existing studies in this area can be classified by four aspects:

- *Centralized or distributed*: For centralized sharing schemes, there exists a central entity in the network who is responsible to make spectrum allocation and access decisions. Sensing procedure can remain distributed, but sensing information has to be aggregated at the central entity. By using centralized schemes, the global optimal access control may be realized. On the contrary, in distributed sharing manners, spectrum allocation and access are based on local polices that are performed by each individual user so that the central controller is no longer needed in this scenario.
- *Cooperative or non-cooperative*: Cooperations may occur among multiple SUs or between PUs and SUs. In the former case, SUs could share interference information so as to achieve a welfare optimality. In the latter, SUs could relay traffics for PUs in exchange for extra radio resources. Non-cooperative sharing may result in a reduced spectrum utilization, while the frequent information exchanges are not required.
- *Overlay or underlay*: Overlay sharing refers to the scenario that SUs can only utilize spectrum that has not been occupied by PUs, while underlay sharing allows SUs to access spectrum that is currently being used by PUs if the interference limit has not been exceeded.
- *Intranetwork or internetwork*: Intranetwork sharing focuses on spectrum allocation among users in a single CR network. On the other hand, internetwork sharing deals with the allocation problem among multiple networks.

In practice, there are plenty of research challenges in realizing efficient spectrum sharing in CR networks as follows:

- *Use of common control channel*: Most of the functionalities in spectrum sharing depend on the use of a perfect common control channel. This is not trivial in CR networks due to the fluctuations of channel availabilities. Consequently, dedicated control channel may be required.
- *Definition of spectrum unit*: Normally, spectrum sharing decisions consider a channel as the basic spectrum unit. Thus, the definition of channels (e.g., characteristics in bandwidth and quality) is essential in designing sharing schemes.

1.2.2.4 Spectrum Mobility

In CR networks, SUs are considered as visitors to temporally and dynamically access the unused licensed spectrum. Thus, if a PU returns and starts accessing a channel which is currently occupying by an SU, the SU needs to vacate the channel and may continue its transmission on another idle channel.

The open research issues in this area are:

- *Spectrum mobility in the time domain*: Since SUs access licensed spectrum based on the spectrum availability, and such availability varies over time, ensuring the QoS of SUs in this environment is always challenging.
- *Spectrum mobility in space*: Naturally, the spectrum availability changes when the SU moves from one place to another. Thus, guaranteeing continuous wireless services for mobile SUs is necessary, but difficult.

1.2.2.5 Models of DSA

Generally, DSA models can be broadly categorized into three classes, i.e., dynamic exclusive use model, open sharing model, and hierarchical access model, as shown in Fig. 1.5.

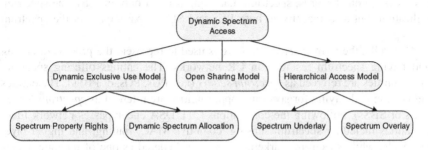

Fig. 1.5 Categories of DSA models

Dynamic exclusive use model maintains the basic structure of current spectrum regulation policy, while introduces flexibility to improve spectrum efficiency. There are two approaches proposed under this model, i.e., spectrum property rights [16] and dynamic spectrum allocation [17]. The former approach is the primitive of spectrum marketing. It enables the licensed holders to lease and trade spectrum by using freely determined technology. The second approach was first raised by European DRiVE project [17], which aims to dynamically assign the spectrum by exploiting spatial and temporal statistics. It can improve the spectrum utilization because it allows much faster variances on spectrum allocation than the current policy. In open sharing model [18], all users are treated as peers for sharing a specific spectrum band. The sharing strategies under this management model have been investigated in both centralized [19] and distributed [20] patterns. Hierarchical access model allows SUs to access unused licensed spectrum while limiting the interference introduced to PUs. Two approaches have been considered: spectrum underlay and spectrum overlay. The underlay approach is based on a worst assumption that PUs transmit all the time. Thus, it constraint the transmission power of SUs so as to make the noise perceived by PUs less than a certain threshold. The advantage of this approach is that it does not need to detect and exploit spectrum holes. Unlike spectrum underlay, overlay sharing requires SUs to identify and use the idle spectrum defined in space, time, and frequency. However, it does not necessarily impose limits on SUs' transmission power.

1.2.3 Market-Driven Dynamic Spectrum Sharing

Most of the existing works in the area of CR-based DSA focus on the traditional framework as illustrated in the previous section. However, the inherent technical challenges of this traditional framework prompt us to find an alternative, which can reduce these limitations so as to be closer to practical implementation. Market-driven dynamic spectrum sharing (i.e., *spectrum marketing*) is one of the promising paradigms to achieve this goal. The objective of spectrum marketing is to maximize the revenue of spectrum owners, and at the same time enhance the satisfaction of CR users. In other words, spectrum marketing can not only increase the spectrum utilization, but also incentivize both PUs and SUs to participate in the spectrum sharing.

Generally, the term *spectrum trading* is used to represent the process of selling and buying spectrum resource in CR networks. The entities offering spectrum opportunities are refereed as *spectrum sellers* (i.e., PUs, POs, or PBSs). The entities demanding and paying for spectrum opportunities are refereed as *spectrum buyers* (SUs or SBSs). Following the conventions [21], DSA encompasses network functionalities including spectrum sensing, cognitive MAC, routing, and higher-layer protocols, while spectrum marketing can be considered as one of its components which deals with economic aspects of DSA. The relationship between DSA and

Fig. 1.6 Relationship between DSA and spectrum marketing

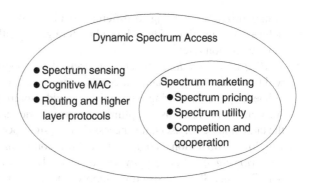

spectrum marketing can be depicted in Fig. 1.6. In order to better understand the market-driven DSA framework, we can partition the spectrum sharing procedure into two major steps, i.e., *spectrum exploration* and *spectrum exploitation*. The objectives of spectrum exploration are to discover the statistics of primary spectrum usage, and identify the spectrum holes. In the spectrum exploitation step, each SU determines the ways to exploit the spectrum opportunities. Based on the specific sharing models, SUs' spectrum access may require explicit permissions from the spectrum owner or cooperation from PUs. Spectrum marketing lies between the spectrum exploration and exploitation steps. Specifically, the spectrum owner first performs spectrum hole identification in the exploration step. After the idle spectrum resources have been determined, they are traded to SUs through spectrum marketing. Finally, SUs proceed to utilize the granted spectrum opportunities in the exploitation step.

Though market-driven DSA can avoid technical challenges, such as sensing in traditional DSA framework, it has its own unique research issues which have to be carefully addressed.

- *Spectrum pricing*: In spectrum marketing, *price* is an essential element which indicates the value of spectrum to both sellers and buyers. For each spectrum buyer, the price paid to the seller depends on its QoS satisfaction through the usage of the certain spectrum bands. For each spectrum seller, the price determines its extra revenue through temporally leasing the idle spectrum resource. The spectrum price should be set based on the spectrum demand from buyers and the spectrum supply of sellers. Besides, the competition among buyers or sellers should also be taken into account.
- *Utility functions*: There are two main kinds of utility functions, one for sellers and the other for buyers. Given the price, the utility function of each seller determines the spectrum supply or transmission parameters (e.g., the number of channels, the number of time slots, or the level of transmission power) for buyers to access the spectrum. This spectrum supply can be derived by maximizing the seller's utility function, which is commonly defined as the difference between revenue and cost from spectrum sharing. The utility function of each buyer implies the amount of spectrum that the buyer demands for a given price so that its satisfaction

is maximized. According to different requirements, the buyer's utility can be defined as a function of its achieved QoS (e.g., logarithmic or sigmoid function of transmission rate).

- *Trading model for large number of users*: In practical spectrum market, there will be a lot of licensed and unlicensed users operating on multiple channels. An efficient spectrum sharing mechanism should capture the behaviors of users in such environment so as to optimize spectrum price and allocation. Because of the large number of users, it is necessary to design spectrum sharing mechanisms with low computational complexity and low communication overhead.

- *Time-varying spectrum supply and demand*: Due to the uncertain activities of both PUs and SUs, the spectrum supply from a seller and the spectrum demand from a buyer can be time varying. With such consideration, adaptive pricing scheme and spectrum access mechanism would be required to achieve the desired system objectives and individual performance guarantee for all users.

References

1. J. Peha, Spectrum management policy options. IEEE Commun. Surv. **1**(1), 2–8, First (1998)
2. A. Goldsmith, *Wireless Communications* (Cambridge University Press, Cambridge, 2005)
3. P. Kolodzy, I. Avoidance, Spectrum policy task force. Federal Communications Commission, Washington, DC, Report ET Docket, no. 02-135, 2002
4. M. Buddhikot, Understanding dynamic spectrum access: models, taxonomy and challenges, in *Proceedings of IEEE DySPAN*, 2007, pp. 649–663
5. Q. Zhao, B. Sadler, A survey of dynamic spectrum access. IEEE Signal Process. Mag. **24**(3), 79–89 (2007)
6. J. Peha, Sharing spectrum through spectrum policy reform and cognitive radio. Proc. IEEE **97**(4), 708–719 (2009)
7. I. Akyildiz, W.-Y. Lee, et al., A survey on spectrum management in cognitive radio networks. IEEE Commun. Mag. **46**(4), 40–48 (2008)
8. J. Mitola, G.Q. Maguire Jr., Cognitive radio: making software radios more personal. IEEE Pers. Commun. **6**(4), 13–18 (1999)
9. S. Haykin, Cognitive radio: brain-empowered wireless communications. IEEE J. Sel. Areas Commun. **23**(2), 201–220 (2005)
10. T. Yucek, H. Arslan, A survey of spectrum sensing algorithms for cognitive radio applications. IEEE Commun. Surv. Tutorials **11**(1), 116–130, First (2009)
11. A. Ghasemi, E. Sousa, Spectrum sensing in cognitive radio networks: requirements, challenges and design trade-offs. IEEE Commun. Mag. **46**(4), 32–39 (2008)
12. I.F. Akyildiz, W.-Y. Lee, et al., Next generation/dynamic spectrum access/cognitive radio wireless networks: a survey. Comput. Netw. **50**(13), 2127–2159 (2006)
13. E. Hossain, D. Niyato, Z. Han, *Dynamic Spectrum Access and Management in Cognitive Radio Networks* (Cambridge University Press, Cambridge, 2009)
14. K. Shin, H. Kim, et al., Cognitive radios for dynamic spectrum access: from concept to reality. IEEE Wirel. Commun. **17**(6), 64–74 (2010)
15. R.V. Prasad, P. Pawelczak, et al., Cognitive functionality in next generation wireless networks: standardization efforts. IEEE Commun. Mag. **46**(4), 72–78 (2008)
16. D. Hatfield, P. Weiser, Property rights in spectrum: taking the next step, in *Proceedings of IEEE DySPAN*, November 2005, pp. 43–55

17. L. Xu, R. Tonjes, T. Paila, et al., DRiVE-ing to the internet: dynamic radio for ip services in vehicular environments, in *Proceedings of IEEE LCN*, 2000, pp. 281–289
18. W. Lehr, J. Crowcroft, Managing shared access to a spectrum commons, in *Proceedings of IEEE DySPAN*, November 2005, pp. 420–444
19. O. Ileri, D. Samardzija, N. Mandayam, Demand responsive pricing and competitive spectrum allocation via a spectrum server, in *Proceedings of IEEE DySPAN*, November 2005, pp. 194–202
20. R. Etkin, A. Parekh, D. Tse, Spectrum sharing for unlicensed bands. IEEE J. Sel. Areas Commun. **25**(3), 517–528 (2007)
21. D. Niyato, E. Hossain, Spectrum trading in cognitive radio networks: a market-equilibrium-based approach. IEEE Wirel. Commun. **15**(6), 71–80 (2008)

Chapter 2
Fundamentals of Mechanism Design

2.1 Introduction

Mechanism design is a subfield of microeconomics and game theory. It considers how to implement good system-wide solutions to problems that involve multiple self-interested agents [1]. In 2007, the Nobel Prize in economics was awarded to Leonid Hurwicz, Eric Maskin, and Roger Myerson "for having laid the foundations of mechanism design theory." This indicates the importance and popularity of mechanism design in various areas of applied economics as well as market-driven applications. For instance, mechanism design has been extensively studied in practical engineering problems, such as electronic market design, distributed scheduling, and radio resource allocation.

Though mechanism design is originated from game theory, it is fundamentally different in terms of design objective and system model. Specifically, traditional game theory emphasizes on the analysis of the outcome of strategic interactions between individual agents in a given game, while mechanism design focuses on designing the game which can produce a certain desired outcome. Thus, mechanism design is also known as *reverse game theory*. A key feature of mechanism design is that the determination of the optimal allocation depends on information which is possessed privately by agents. In order to obtain the optimal solution, this private information has to be elicited from agents. However, agents may be sophisticated and intelligent to misreport their private information rather than truthful telling if they can recognize the potential benefit from such behaviors. Apparently, computing the allocation from incorrect information may result in serious mistakes. Hence, it is challenging to devise a mechanism or a procedure for information interactions such that the outcome is optimal even when these agents behave strategically. Mechanism design can therefore be considered as a design of rules which can guarantee desirable outcomes amongst fully strategic agents [2].

© The Author(s) 2016
C. Yi, J. Cai, *Market-Driven Spectrum Sharing in Cognitive Radio*,
SpringerBriefs in Electrical and Computer Engineering,
DOI 10.1007/978-3-319-29691-3_2

In mechanism design systems, there are two main kinds of components [3]. They are the apparatus under the control of the designer, called *mechanism*, and the world of things that are beyond the designer's control, called *environment*. A mechanism consists of rules that govern what the participants are permitted to do and how these permitted actions determine outcomes. An environment comprises the list of potential participants, participants' private types (e.g., preferences, information, and beliefs), and all possible outcomes. For example, a single-item selling mechanism $(\boldsymbol{b}, \boldsymbol{\pi}, \boldsymbol{\mu})$ has the following components: a set of types $\boldsymbol{b}$, an allocation rule $\boldsymbol{\pi}$, and a payment rule $\boldsymbol{\mu}$ for all buyers. As functions of all buyers' types, the allocation and payment rules determine the probability $\pi_i(\boldsymbol{b})$ that buyer i will obtain the item and the expected payment $\mu_i(\boldsymbol{b})$ that buyer i has to make, respectively.

Auction is one of the most typical and well-known pricing-based mechanisms, which aims to allocate resources among a large number of bidders. To be more specific, an auction mechanism is a process of resource allocation and price discovery on the basis of bids from participants. Following the general structure of mechanisms, an auction model includes three major parts: a description of the potential bidders, the set of possible resource allocations (describing the number of goods, whether the goods are divisible, and whether there are any restrictions on how the goods should be allocated), and the values of different resource allocations to each bidder. The mechanism designer's problem is to determine the rules of the auction, i.e., which bids can be allowed, how the resources are allocated, and what prices should be charged, to achieve some objectives, such as maximizing the social welfare or seller's revenue. Due to the broad adaptability of auction mechanisms and the natural market relationship between spectrum owners and buyers in CR networks, auction mechanisms have been recently considered as an effective and promising approach for the future realization of dynamic spectrum sharing [4].

In the rest of this chapter, we particularly focus on auction mechanism design. Specifically, the basic structure and design objectives of auction mechanism are reviewed. After that, three important auction mechanisms are introduced and analyzed in detail in order to facilitate readers to understand the advanced spectrum auction mechanisms in Chaps. 3, 4, and 5. Finally, a concise literature survey is provided to summarize the existing works of mechanism design in spectrum sharing problems.

2.2 Auction Mechanism Design

Auction is an incentive approach to encourage all users to participate in the market by choosing their own strategies while guaranteeing certain design goals. In other words, an auction mechanism motivates users to make bids based on their own individual rationalities, and then the desired auction properties and objectives will be naturally achieved as expected by executing the mechanism based on the designed rules. As a necessary prerequisite, some terminologies and general concepts in auction mechanism design are first presented as follows [5].

- *Mechanism and strategy*: A mechanism commonly consists of two steps. The first step is the strategy submission (e.g., bidding) from all the participants (i.e., buyers and sellers). Each participant has a type which indicates its own private preference, and will affect the participant's strategies. The second step is the outcome determination of the system. Specifically, in an auction mechanism, each buyer (or seller) has private information on the auction commodities. The strategies (i.e., bids from buyers or asks from sellers) will be submitted to the auctioneer, each of which reflects the participant's preference. Then, the auctioneer runs the mechanism and determines the optimal outcome based on a set of pre-determined auction rules. When applying auctions to practical systems, it may be feasible to directly adopt the existing auction mechanisms or re-design new auction mechanisms which can maintain basic properties (e.g., economic robustness) and satisfy the requirements of all participants.
- *Utility, revenue, social welfare and incentive design*: In auctions, the buyer who receives its requested commodities has a utility which equals the difference between its valuation (i.e., a function of its type) and the final hammer price. The seller who sells the commodities can obtain a revenue which is the gap between the hammer price and its own valuation. Clearly, revenue is also the utility of the seller. Other buyers (sellers) will receive zero utilities (revenues) throughout the auction. The sum of all buyers' utilities and sellers' revenues is defined as the social welfare, which represents the profit that the mechanism produces to the market. Non-negative utilities and revenues indicate extra economic benefits, which can be seen as a kind of incentive to encourage more participants to join the auction. Any auction mechanism should be designed to provide enough incentive to all participants. Failing to do so will lead to a collapse of the market since fewer and fewer buyers (sellers) will be interested in the auction, and commodities cannot be efficiently allocated [6].

2.2.1 Basics of Auction Mechanism

An auction mechanism can be viewed as a process of buying and selling commodities or services. Generally, an auction consists of the following basic elements:

- *Buyer*: The one who wants to buy commodities in auctions is regarded as a buyer. For wireless communications, buyers are users who are eager to obtain radio resources for their own transmissions through pricing competitions with other users.
- *Seller*: As another kind of players in the auction, sellers own commodities and are willing to sell them for potential economic profits. In spectrum auctions, sellers could be any spectrum holders, e.g., the regulator (FCC), POs, PUs, or PBS.
- *Auctioneer*: An auctioneer acts as an intermediate agent and a central controller who hosts and runs auction processes between sellers' and buyers' sides.

In general, auctioneers could be non-profit entities, third-party brokers, or even the sellers themselves. For instance, a base station or an access point in wireless networks can conduct its own radio resource auctions.

- *Commodity*: In the market, commodities are also known as goods which can be traded between sellers and buyers. In radio resource auctions, such commodity could be spectrum bandwidth, licenses of spectrum, time slots, and transmission power levels.
- *Valuation*: Valuations represent the monetary evaluation of assets. Every buyer should have its own valuation towards its demand. However, different buyers may have different valuations for the same commodity due to their personal preferences (i.e., types). Valuations can be *private*, which means that buyers do not know the valuations of each other, or *public* so that such information is known to all the others. Note that the scenarios with completely public valuations are commonly known as *open-cry auctions*, which are beyond the scope of conventional mechanism design, and thus will not be discussed in this brief.
- *Price*: During an auction, a seller can submit an *ask* to indicate an asking price on its selling commodities. Sometimes, asking prices are not necessary (i.e., equal to zero), e.g., spectrum of PUs may have no value if it remains idle. On the other hand, a buyer can submit a *bid* to inform the bidding price for its demanded commodities. A *hammer price* is determined by the auctioneer, indicating the final payments of buyers and earnings of sellers.

Numerous auction mechanisms have been proposed and applied in practical markets. From the view of auction formats, we summarize some typical ways of categorization, which are widely discussed in the literature.

- *Forward or reverse*: In forward auctions, buyers compete by bidding for commodities from seller(s), as shown in Fig. 2.1a. On the contrary, in reverse auctions, sellers compete to sell commodities to buyers instead, as shown in

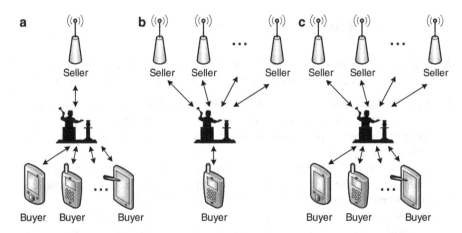

Fig. 2.1 Illustration of different auction formats. (**a**) Forward auction. (**b**) Reverse auction. (**c**) Double-sided auction

Fig. 2.1b. Generally, if the objective of a forward auction is to maximize the social welfare, then the corresponding objective of the reverse auction is to minimize the welfare.

- *Single-sided or double-sided*: Single-sided auction refers to the case with only buyers or sellers competing in the auction (both scenarios shown in Fig. 2.1a, b). If competitions exist in both sellers' and buyers' sides, the problem is formulated as a double-sided auction, as illustrated in Fig. 2.1c. For single-sided auctions, the auctioneer may not be required since either the seller (in forward auctions) or the buyer (in reverse auctions) could concurrently play the role as the auctioneer. However, it is necessary to have an auctioneer in double-sided auctions to collect asks and bids from sellers and buyers, respectively, and matches those prices by allocating commodities from sellers to buyers, as well as payments from buyers to sellers accordingly.
- *Single-unit or multi-unit*: Both the demand from buyers and the supply from sellers could be either single-unit or multi-unit. The buyer (seller) with single-unit demand (supply) can only bid (provide) one commodity at a time, while the buyer (seller) with multi-unit demand (supply) can bid (provide) multiple commodities simultaneously.
- *Offline or online*: In offline auctions, buyers (sellers) are allowed to make bids (asks) within the bidding period, and the market is only cleared at a certain specified time instants after the bidding. However, in online auctions, whenever the asks and the bids arrive, the market is cleared with immediate allocation and payment decisions.

2.2.2 Design Goals and Properties

In order to understand the core of auction mechanisms and the key of applying auctions in market-driven applications, we formally introduce the essential objectives and properties of auction mechanism design.

For illustration purpose and the convenience of explanation, let us consider a forward auction system with a single seller who wants to sell an indivisible commodity to N potential buyers. Each buyer i has a private value (i.e., type) v_i over the commodity, and bids b_i in the auction. The vectors $\boldsymbol{v} = \{v_1, \ldots, v_N\}$ and $\boldsymbol{b} = \{b_1, \ldots, b_N\}$ represent the sets of all buyers' values and bids, respectively. For simplicity, the value of the commodity to the seller is assumed to be 0. First, we recall some general definitions in traditional game theory [7] to describe the auction mechanism:

Definition 2.1 (Dominant Strategy). A dominant strategy of a buyer is one that maximizes its utility regardless of other buyers' strategies. Specifically, b_i is the dominant bidding strategy for buyer i if its utility U_i cannot be improved for any $b_i' \neq b_i$, and any strategy profile of the other buyers $\boldsymbol{b}_{-i} = \{b_1, \ldots, b_{i-1}, b_{i+1}, \ldots, b_N\}$, i.e.,

$$U_i(b_i, \boldsymbol{b}_{-i}) \geq U_i(b_i', \boldsymbol{b}_{-i}). \tag{2.1}$$

Fig. 2.2 Revelation principle

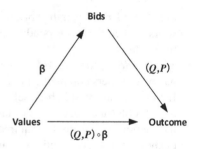

Definition 2.2 (Equilibrium). An N-tuple of bidding strategies $b := \beta = \{\beta_1(v_1), \ldots, \beta_N(v_N)\}$ is an equilibrium of the auction mechanism if for every buyer i, given the strategies β_{-i} of all other buyers, $\beta_i(v_i)$ is the dominant strategy of buyer i that maximizes its utility.

Since there is no restrictions on bids, an auction mechanism may be very complicated due to the countless combinations of arbitrary bidding strategies. A smaller and simpler class of mechanisms is the one in which all buyers bid by directly revealing their own private values, i.e., $b_i = v_i$ for every buyer i. Such mechanisms are called *direct mechanisms* [8].

Definition 2.3 (Direct Mechanism). All buyers are asked to directly report their values v in the direct mechanism (Q, P), where $Q = \{Q_1(v), \ldots, Q_N(v)\}$ and $P = \{P_1(v), \ldots, P_N(v)\}$. As the outcome, $Q_i(v) \in \{0, 1\}$ and $P_i(v) \in \mathbb{R}^+$ indicate the allocation decision and the payment for buyer i, respectively.

Since the direct mechanism is possible to be implemented only if the truthful revelation is an equilibrium for all buyers, we have to explore the relationship between direct mechanisms and auction equilibriums. A well known result is the *revelation principle* [8–10], which shows that the outcomes resulting from any equilibrium of any auction mechanism can be replaced by the outcome of a direct mechanism. In this sense, there is no loss of generality to limit our focus on direct mechanisms.

Theorem 2.1 (Revelation Principle). *Given an auction mechanism with a corresponding equilibrium, there exists a direct mechanism in which (i) it is an equilibrium for each buyer to report its value truthfully, and (ii) the outcome is the same as the one produced by the equilibrium of the given mechanism.*

Proof. From definitions, we have $Q = Q(\beta(v))$ and $P = P(\beta(v))$, which demonstrates that the direct mechanism (Q, P) is a composition of functions (Q, P) and β, as shown in Fig. 2.2. Thus, conclusions (i) and (ii) can be verified routinely. $\square$

The underlying idea of revelation principle can be interpreted as follows: Fix a mechanism and the corresponding equilibrium β of the auction. Instead of having each buyer i submit bids $b_i = \beta_i(v_i)$ and then applying the rules in this mechanism to determine the outcome (i.e., the allocations and payments), we can ask all buyers to directly report their truthful values while guarantee that the outcome is the same as the case when they bid β. In other words, the direct mechanism "calculates" the equilibrium for buyers automatically.

In the following, some desired properties of auction mechanisms with direct revelation are presented.

Definition 2.4 (Incentive Compatibility or Truthfulness). An auction mechanism (Q, P) is incentive-compatible or truthful if no buyer i can improve its own utility U_i by bidding $b_i = v_i' \neq v_i$, i.e.,

$$U_i := v_i \cdot Q_i(v_i) - P_i(v_i) \geq v_i' \cdot Q_i(v_i') - P_i(v_i'), \qquad \forall i. \qquad (2.2)$$

Definition 2.5 (Individual Rationality). An auction mechanism (Q, P) is individual-rational if every buyer i can receive a non-negative utility after the auction, i.e., $U_i \geq 0$.

Incentive compatibility is essential to resist market manipulation and ensure auction fairness and efficiency [9]. In untruthful auctions, selfish buyers may misreport their values to manipulate the system and obtain outcomes that favor themselves but hurt the others. While in truthful auctions, the dominant strategy for each buyer is to bid truthfully so that the possibility of market manipulation and the overhead of strategizing over others are eliminated. Besides, *individual rationality* can ensure economic benefits for all participants, and thus encourage all users to join the auction. Clearly, an economic robust auction mechanism should guarantee both the properties of incentive compatibility and individual rationality.

2.2.3 Second-Price Sealed-Bid Mechanism

Second-price sealed-bid (SPSB) mechanism is also well-known as *Vickrey auction mechanism* [11] which was first proposed by the Nobel prize winner Dr. William Vickrey in 1961. In this mechanism, buyers are asked to submit sealed bids for a single commodity. The buyer who reports the highest bid wins (i.e., it is awarded the commodity), while pays the price of the second highest bid.

Let v_i and b_i be the value and bid of each buyer i in the auction, respectively. Then, the utility of buyer i can be expressed as

$$U_i = \begin{cases} v_i - \max_{j \neq i} b_j, & \text{if } b_i > \max_{j \neq i} b_j; \\ 0, & \text{if } b_i < \max_{j \neq i} b_j. \end{cases} \qquad (2.3)$$

It is also assumed that if there is a tie, i.e., $b_i = \max_{j \neq i} b_j$, for any buyer i, the commodity goes to any of these buyers with an equal probability. Since SPSB mechanism apparently guarantees the individual rationality (i.e., $U_i \geq 0$) through its payment rule, only the incentive compatibility remains to be examined.

Proposition 2.1. *SPSB mechanism is incentive-compatible, i.e., it is a dominant strategy for every buyer to bid by directly revealing its value in the auction.*

Proof. We consider the scenario with potential overbidding of each buyer i (i.e., $b_i > v_i$). If $\max_{j \neq i} b_j < v_i$, buyer i will win the auction with a truthful bid ($b_i = v_i$) as well as an overbid. The bidding price cannot change its obtained utility in this case since the payment remains unchanged as $\max_{j \neq i} b_j$. If $\max_{j \neq i} b_j > b_i$, the bidder will lose by either bidding truthfully or untruthfully, so that the buyer's utility will be 0 at the end for both cases. If $b_i > \max_{j \neq i} b_j > v_i$, then only the strategy of overbidding will win the auction. However, the utility will be negative by overbidding because the buyer has to pay more than its value, while the utility for truthful bidding is 0. Hence, the outcome of overbidding is never better than the one with truthful bids. A similar argument shows that it is not profitable to bid less than v_i. In conclusion, truthful bidding dominates the other possible strategies (underbidding and overbidding), and thus SPSB mechanism is incentive-compatible.   □

2.2.4 Vickrey–Clarke–Groves Mechanism

Vickrey–Clarke–Groves (VCG) mechanism [11–13] is a generalized version of SPSB mechanism. It aims to assign multiple commodities to self-interested buyers in a socially optimal manner. In other words, VCG mechanism charges each individual buyer the "harm" it causes to other buyers [8], and ensures that the optimal bidding strategy for each buyer is to reveal its true value. In addition, VCG mechanism is a basic but effective tool for dealing with combinatorial auction problems [14].

In order to specify the details of VCG mechanism, let us take the single-minded[1] combinatorial auction as an example. Consider a seller who wants to auction off a set $\mathcal{S}$ of commodities among N buyers with heterogeneous demands, where any subset $\mathcal{T} \subseteq \mathcal{S}$ is called as a *bundle*. Each buyer i has a value v_i which is a function of its requested bundle $\mathcal{T}_i \subseteq \mathcal{S}$, and thus such value can also be denoted as $v_i(\mathcal{T}_i)$. Different from the single-item auction, there can be multiple winners in a combinatorial auction, and the allocation decision (winner determination) is to decide whether grant each buyer i with its demand bundle $\mathcal{T}_i$. In addition, winners can be charged by different payments in combinatorial auctions. Without loss of generality, the utility of each buyer i can be written as

$$U_i = v_i(\mathcal{T}_i) \cdot x_i - p_i, \tag{2.4}$$

where $x_i = 0$ or 1 indicates losing or winning of buyer i, and $p_i \in \mathbb{R}^+$ is its final payment.

---

[1]This refers to win-or-lose scenarios where each user is only interested in getting all it demands or nothing.

The procedures of the corresponding VCG mechanism can be described as follows:

(1) Each buyer submits a bid $b_i(\mathcal{T}_i)$ for its demanding bundle $\mathcal{T}_i \subseteq \mathcal{S}$. Note that, $b_i(\mathcal{T}_i) = v_i(\mathcal{T}_i)$ if the buyer behaves truthfully.
(2) The auctioneer (or the seller itself) determines a feasible allocation and a winner assignment that maximizes the total bidding price from all winners. Note that this is equivalent to social welfare maximization since the seller's revenue and buyers' total payments can be cancelled out. Thus, for all feasible winner assignments $\{x_i\}_{i=1}^N$,[2] the objective is

$$\max_{\{x_i\}_{i=1}^N} \sum_{i=1}^N b_i(\mathcal{T}_i) \cdot x_i. \tag{2.5}$$

(3) With the optimal winner determination $\{x_i^*\}_{i=1}^N$ from (2.5), charge each buyer i with an appropriate price p_i which is calculated as

$$p_i = \left(\max_{\{x_j\}_{j \neq i}} \sum_{j \neq i} b_j(\mathcal{T}_j) \cdot x_j \right) - \sum_{j \neq i} b_j(\mathcal{T}_j) \cdot x_j^*. \tag{2.6}$$

In (2.6), the first term is the maximum possible welfare for the auction without the participation of buyer i. Obviously, this term can be obtained by removing the bid of buyer i from the input and optimizing the objective in (2.5) for the rest $N - 1$ buyers. The second term collects all bids from the optimal allocation $\{x_i^*\}_{i=1}^N$ except for buyer i. Therefore, this payment reflects the degradation (i.e., the harm) on the social welfare of all other buyers due to the presence of buyer i.

Proposition 2.2. *The VCG mechanism is economically efficient which means that, if all buyers bid truthfully, then the VCG mechanism outputs an allocation that maximizes the social welfare over all feasible allocations.*

Proof. This can be directly observed from step (2) of the procedure. □

Proposition 2.3 (Individual Rationality). *VCG mechanism guarantees that the utility of any truthful buyer is always non-negative.*

Proof. Given the payment function in (2.6) and $b_i = v_i$, we can add and subtract a term, $v_i(\mathcal{T}_i) \cdot x_i^*$, so that p_i can be rewritten as

---

[2]"Feasible" means that for any two winners $i \neq j$, we have $\mathcal{T}_i \cap \mathcal{T}_j = \emptyset$.

$$p_i = v_i(T_i) \cdot x_i^* - v_i(T_i) \cdot x_i^* + \left(\max_{\{x_j\}_{j \neq i}} \sum_{j \neq i} v_j(T_j) \cdot x_j \right) - \sum_{j \neq i} v_j(T_j) \cdot x_j^*$$

$$(2.7)$$

$$= v_i(T_i) \cdot x_i^* - \left[\sum_{j=1}^{n} v_j(T_j) \cdot x_j^* - \left(\max_{\{x_j\}_{j \neq i}} \sum_{j \neq i} v_j(T_j) \cdot x_j \right) \right]$$

Hence, the proof of this proposition is to show that the discount term in (2.7) is non-negative since $U_i = v_i(T_i)x_i^* - p_i$. It holds because adding an extra buyer will never decrease the maximum achievable welfare (this leads to a larger set of feasible allocations). □

Proposition 2.4 (Incentive Compatibility). *VCG mechanism guarantees that bidding truthfully is a dominant strategy for every buyer in the auction.*

Proof. To prove the incentive compatibility, we need to show that for every buyer i with value v_i, its utility when bidding v_i is not less than the utility when bidding $v_i' \neq v_i$. Let $\boldsymbol{b} = \{v_i, \boldsymbol{v}_{-i}\}$ and $\boldsymbol{b}' = \{v_i', \boldsymbol{v}_{-i}\}$, where $\boldsymbol{v}_{-i} = \{v_1, \dots, v_{i-1}, v_{i+1}, \dots, v_N\}$. Clearly, the outcome of the VCG mechanism depends on bids from all buyers. Thus, the utility of buyer i when declaring v_i in VCG mechanism can be expressed as

$$U_i(v_i, \boldsymbol{v}_{-i}) = v_i \cdot x_i^*(\boldsymbol{b}) - \left[\left(\max_{\{x_j\}_{j \neq i}} \sum_{j \neq i} v_j \cdot x_j(v_j) \right) - \sum_{j \neq i} v_j \cdot x_j^*(\boldsymbol{b}) \right]$$

$$(2.8)$$

$$= \sum_{i=1}^{N} v_i \cdot x_i^*(\boldsymbol{b}) - \max_{\{x_j\}_{j \neq i}} \sum_{j \neq i} v_j \cdot x_j(v_j),$$

while the utility of buyer i when declaring v_i' is

$$U_i(v_i', \boldsymbol{v}_{-i}) = v_i \cdot x_i^*(\boldsymbol{b}') - \left[\left(\max_{\{x_j\}_{j \neq i}} \sum_{j \neq i} v_j \cdot x_j(v_j) \right) - \sum_{j \neq i} v_j \cdot x_j^*(\boldsymbol{b}') \right]$$

$$(2.9)$$

$$= \sum_{i=1}^{N} v_i \cdot x_i^*(\boldsymbol{b}') - \max_{\{x_j\}_{j \neq i}} \sum_{j \neq i} v_j \cdot x_j(v_j).$$

Subtracting (2.8) by (2.9), we have

$$U_i(v_i, \boldsymbol{v}_{-i}) - U_i(v_i', \boldsymbol{v}_{-i}) = \sum_{i=1}^{N} v_i \cdot x_i^*(\boldsymbol{b}) - \sum_{i=1}^{N} v_i \cdot x_i^*(\boldsymbol{b}') \geq 0. \qquad (2.10)$$

The second inequality holds since the definition of $\{x_i^*\}_{i=1}^N$ implies that the social welfare (i.e., $\sum_{i=1}^{N} v_i \cdot x_i^*$) is maximized by truthful telling. □

2.2.5 Lehmann–Oćallaghan–Shoham Mechanism

A major drawback of VCG mechanism is that it highly relies on the optimal solution of the winner determination problem (WDP). In practice, WDPs are normally Non-deterministic Polynomial-time hard (NP-hard), especially for combinatorial auctions. Thus, one possible way is to implement approximate-optimal algorithms for solving WDPs in polynomial time. Since VCG mechanism is incompatible with any approximate allocations [15], we introduce another powerful and polynomial-time approximate approach, called *Lehmann–Oćallaghan–Shoham (LOS) Mechanism* [16]. This mechanism consists of a *LOS WDP algorithm* and a *LOS payment design*.

Following the same example of combinatorial auction described in Sect. 2.2.4, we consider that the seller has m commodities to sell, and each buyer i has a demand T_i and a bidding price b_i. The procedure of LOS WDP algorithm is summarized as follows:

(1) The auctioneer (or the seller itself) re-indexes the received bids such that

$$\frac{b_1}{\sqrt{|T_1|}} \geq \frac{b_2}{\sqrt{|T_2|}} \geq \cdots \geq \frac{b_n}{\sqrt{|T_n|}}. \tag{2.11}$$

(2) Based on the above order, check buyer i from 1 to N: if no commodity in T_i has already been assigned to previous buyers, set $x_i = 1$ to indicate that buyer i becomes a winner; otherwise, set $x_i = 0$ to indicate that buyer i loses the auction.

Proposition 2.5. *LOS WDP algorithm is a $\sqrt{m}$-approximation algorithm, where m is the number of commodities.*

Proof. Let W and W^* denote the sets of winners granted by the LOS and optimal WDP algorithms, respectively. We need to prove that

$$\sum_{i^* \in W^*} b_{i^*} \leq \sqrt{m} \cdot \sum_{i \in W} b_i. \tag{2.12}$$

We first make a simple but crucial definition. We say that a buyer $i \in W$ *blocks* a buyer $i^* \in W^*$ if $T_i \cap T_j \neq \emptyset$. Note that $i = i^*$ is allowed in this definition, and demands T_i and T_{i^*} cannot both be granted if buyer i blocks buyer i^* and $i \neq i^*$. For a buyer $i \in W$, let $F_i \subseteq W^*$ denote the set of buyers in W^* that is *first* blocked by buyer i (i.e., $i^* \in W^*$ is placed in F_i if and only if i is the first bid in the greedy ordering (2.11) that blocks i^*).

There are two key points. First, suppose $i^* \in F_i$ (i.e., $i^* \in W^*$ is first blocked by $i \in W$). Then, at the time that LOS WDP algorithm chose to grant the request from buyer i, buyer i^* was not yet blocked and was a viable alternative. Thus, by (2.11), we must have

$$\frac{b_i}{\sqrt{|T_i|}} \geq \frac{b_{i^*}}{\sqrt{|T_{i^*}|}}, \quad \forall i^* \in F_i. \tag{2.13}$$

The second key point is that each optimal $i^* \in W^*$ lies in precisely one set F_i. In other words, each buyer $i^* \in W^*$ must be blocked by at least one buyer in W (possibly by itself), since buyer i^* would only be passed over by the greedy algorithm if it was blocked by some previously granted buyers. Thus, F_i's are a partition of W^* and we have

$$\sum_{i^* \in W^*} b_{i^*} = \sum_{i \in W} \sum_{i^* \in F_i} b_{i^*}. \tag{2.14}$$

This fact allows us to consider each bid separately and then combine those results to obtain a global bound.

Now, consider a buyer $i \in W$. Summing over all $i^* \in F_i$ in (2.13), we have

$$\sum_{i^* \in F_i} b_{i^*} \leq \frac{b_i}{\sqrt{|T_i|}} \sum_{i^* \in F_i} \sqrt{|T_{i^*}|}. \tag{2.15}$$

In addition, since all buyers in F_i were simultaneously granted by the optimal solution, their demands must be disjoint and hence

$$\sum_{i^* \in F_i} |T_{i^*}| \leq m. \tag{2.16}$$

Moreover, with (2.15) and (2.15), applying Cauchy–Schwarz inequality [17] gives

$$\sum_{i^* \in F_i} b_{i^*} \leq \frac{b_i}{\sqrt{|T_i|}} \sum_{i^* \in F_i} \sqrt{\frac{m}{|F_i|}} = \sqrt{m} \cdot \frac{b_i}{\sqrt{|T_i|}} \sqrt{|F_i|}. \tag{2.17}$$

Furthermore, since buyer i blocks all of the buyers in F_i, and demands in F_i are disjoint, we have $|F_i| \leq |T_i|$, which implies that

$$\sum_{i^* \in F_i} b_{i^*} \leq \sqrt{m} \cdot b_i. \tag{2.18}$$

Finally, summing over all $i \in W$ and applying (2.14), we can observe that the inequality (2.12) holds.   □

The basic idea of LOS payment design is to charge each buyer with the price that are "Vickrey-like." Before presenting the detailed payment scheme, we introduce the definition of u-blocks.

Definition 2.6 (u-Blocks). By applying LOS WDP algorithm, suppose that buyer i wins while buyer j loses. If buyer j could win by removing the bid of buyer i from the input of LOS WDP algorithm, we say that buyer i u-blocks buyer j.

In LOS payment scheme, a winning buyer will be charged by the "most valuable" bid from buyers it u-blocks. Specifically, this payment scheme can be demonstrated as follows:

- If buyer i loses or wins but u-blocks no other buyer, then its payment is set as $p_i = 0$.
- If buyer i is granted with its demand $\mathcal{T}_i$ by the WDP algorithm, and let buyer j (with $\mathcal{T}_j$ and b_j) be the one with the lowest index in (2.11) that buyer i u-blocks, then the payment of buyer i is set as

$$p_i = \frac{b_j}{\sqrt{|\mathcal{T}_j|}} \cdot \sqrt{|\mathcal{T}_i|}. \tag{2.19}$$

Proposition 2.6 (Individual Rationality). *LOS mechanism guarantees that the utility of any truthful buyer is always non-negative.*

Proof. There are three possible outcomes for buyers in LOS mechanism: (1) For losing buyers, their utilities are always 0 since they do not need to make any payments; (2) For winning buyers without u-blocks, their utilities are always positive since they do not need to make any payments either; and (3) For buyer i with $b_i = v_i$ and $\mathcal{T}_i$, let buyer j (with $b_j = v_j$ and $\mathcal{T}_j$) be the one with the lowest index in (2.11) that buyer i u-blocks. According to the ordering rule in LOS WDP algorithm, we must have

$$\frac{v_i}{\sqrt{|\mathcal{T}_i|}} \geq \frac{v_j}{\sqrt{|\mathcal{T}_j|}}. \tag{2.20}$$

Thus,

$$U_i = v_i - p_i = v_i - \frac{v_j}{\sqrt{|\mathcal{T}_j|}} \cdot |\mathcal{T}_i| \geq v_i - \frac{v_i}{\sqrt{|\mathcal{T}_i|}} \cdot |\mathcal{T}_i| = 0. \tag{2.21}$$

In conclusion, LOS mechanism is individual-rational. $\square$

Proposition 2.7 (Incentive Compatibility). *LOS mechanism guarantees that bidding truthfully is a dominant strategy for every buyer in the auction.*

Proof. We can first assume by contradiction that there is a buyer i and a set of bids (including both demand and bidding price) $\{(\mathcal{T}_j, b_j)\}_{j \neq i}$ for all the other $N-1$ buyers such that buyer i can obtain a strictly larger utility by untruthful bidding $b_i \neq v_i$. Define two sets $B_{-i} = \{(\mathcal{T}_j, b_j)\}_{j \neq i}$, $B_T = B_{-i} \cup \{(\mathcal{T}_i, v_i)\}$ and $B_F = B_{-i} \cup \{(\mathcal{T}_i, b_i)\}$. By Proposition 2.6, we can assume that bid $(\mathcal{T}_i, b_i)$ is granted by LOS mechanism given the input B_F. Otherwise, any untruthful behaviors will only make buyers lose the auction, which ends the proof immediately.

Now, consider the case where $b_i < v_i$. Since $(\mathcal{T}_i, b_i)$ can be granted by LOS mechanism and $v_i > b_i$, the bid $(\mathcal{T}_i, v_i)$ would have been considered earlier in the LOS ordering, and thus would also have been granted. Our remaining work is to prove that the payment for untruthful bidding is always higher than the one by truthful bidding. Suppose that $(\mathcal{T}_j, b_j)$ is the first bid u-blocked by $(\mathcal{T}_i, b_i)$. We can complete the proof by showing that the bid $(\mathcal{T}_i, v_i)$ does not u-block any bid earlier than $(\mathcal{T}_j, b_j)$, as then the payment charged by LOS mechanism for $(\mathcal{T}_i, v_i)$ cannot be higher than the one for $(\mathcal{T}_i, b_i)$.

Again by contradiction, we can assume that the first bid $(\mathcal{T}_k, b_k)$ that $(\mathcal{T}_i, v_i)$ u-blocks has lower index than $(\mathcal{T}_j, b_j)$ in the LOS ordering. According to the definition of u-blocking, $(\mathcal{T}_k, b_k)$ can be granted if the bid $(\mathcal{T}_i, v_i)$ is removed from the input B_T of the LOS mechanism. A key observation is that if $(\mathcal{T}_i, b_i)$ follows after $(\mathcal{T}_k, b_k)$ in the LOS ordering, $(\mathcal{T}_k, b_k)$ would also be granted by LOS mechanism given input B_F. Since $\mathcal{T}_i$ and $\mathcal{T}_k$ must have at least one demanded commodity in common, and $(\mathcal{T}_i, b_i)$ can be granted by the LOS mechanism, it implies that $(\mathcal{T}_i, b_i)$ has lower index that $(\mathcal{T}_k, b_k)$ in the LOS ordering. But then $(\mathcal{T}_i, b_i)$ will definitely u-block $(\mathcal{T}_k, b_k)$, which indicates that $(\mathcal{T}_j, b_j)$ has to be in front of $(\mathcal{T}_k, b_k)$ in order to be firstly u-blocked by $(\mathcal{T}_i, b_i)$. However, this clearly contradicts our assumption, and thus proves that bidding $b_i < v_i$ will not increase the utility of any buyer i. A similar argument can be obtained for the case where $b_i > v_i$. Hence, in conclusion, LOS mechanism is incentive-compatible.   □

2.3 Applications of Mechanism Design in Spectrum Sharing

To bridge the gap between mechanism design theory and the practical spectrum sharing problems in wireless communications, we conduct a concise survey on the existing literature with specific focus on auction-based spectrum sharing mechanisms. For convenience of reading, the following discussions are classified into three main categories, i.e., *single-sided mechanisms*, *double-sided mechanisms*, and *online mechanisms*.

2.3.1 Single-Sided Spectrum Sharing Mechanisms

Single-sided mechanism aims to either *buyer-sided auction* with multiple competitive buyers and a single seller or *seller-sided auction* with multiple competitive sellers and a single buyer. In these scenarios, the role of the auctioneer can be integrated in the single seller (buyer) for buyer-sided (seller-sided) auctions since the competition only exists at buyers' (sellers') side.

Since dynamic spectrum sharing models commonly consist of a large number of secondary spectrum buyers competing for accessing the primary licensed spectrum bands, buyer-sided auction mechanism is a natural fit and has been widely discussed in existing works. For instance, Huang et al. in [18] analyzed an underlaying spectrum sharing mechanism among a group of users, subject to a constraint of a certain interference temperature. In [19], Kash et al. proposed a truthful and scalable auction mechanism which aimed to allocate spectrum to both sharers and exclusive-users in secondary networks. Gao et al. in [20] introduced an integrated contract and auction mechanism to maximize the PO's expected profit under stochastic information of CR networks. In [21], Lim et al. studied a cooperation-based dynamic spectrum leasing mechanism via multi-winner auction of multiple bands. Zhan et al. in [22] explored short-interval secondary spectrum markets, where secondary spectrum buyers are allowed to demand flexible number of channels from the single PO. Chen et al. in [23] designed a truthful spectrum auction framework in which buyers could bid and obtain spectrum with variable bandwidth. A fully distributed auction algorithm was studied in [24], which aimed to allocate sub-bands to users so as to maximize the sum-rate of the system. Zheng et al. in [25] discussed an incentive-compatible combinatorial auction for heterogeneous channel allocation with channel spatial reusability. In [26], Li et al. presented a truthful auction mechanism in which the spectrum bandwidth was allocated in a time-frequency division manner. Wu et al. in [27] proposed a privacy-preserving and truthful spectrum auction mechanism which guaranteed anonymity for both single- and multi-channel auctions.

Seller-sided auction mechanism, on the other hand, is less intuitive to be applied in dynamic spectrum sharing. However, it has been recently considered as an effective market-driven mechanism for cellular traffic offloading [28] and mobile crowdsourcing [29]. For example, Zhou et al. in [30] provided a novel incentive mechanism to motivate mobile users to leverage their delay tolerance for cellular traffic offloading. In [31], Dong et al. investigated a novel auction-based incentive framework that allowed a cellular service provider to buy capacities from third-party owners whenever needed through reverse auction mechanism. Koutsopoulos in [32] matched the crowdsourcing problem to the optimal auction mechanism design [33] by assuming that the cost information of users followed a known distribution. In [34], Feng et al. designed a truthful auction mechanism for crowdsourcing with location awareness and coverage.

2.3.2 Double-Sided Spectrum Sharing Mechanisms

Double-sided mechanism aims to the scenarios with competitions on both sellers' and buyers' sides, and thus provides a possible way to model the spectrum allocations from multiple primary spectrum sellers to multiple secondary spectrum buyers. In this mechanism, both sellers and buyers report their prices (i.e., asks and bids) for the trading resources, and the auctioneer collects the auction information, matches the bids and asks, and determines the payments and payoffs for buyers and sellers, respectively.

Zhou et al. in [35] first proposed a general framework for truthful double spectrum auction mechanisms, where multiple parties could trade spectrum based on their individual needs. In [36], a truthful double auction mechanism was studied for heterogeneous spectrum where the distinctive characteristics in both spatial and frequency domains were considered. Gao et al. in [37] analyzed a multi-auctioneer progressive auction mechanism, in which each auctioneer (i.e., seller) systematically raised the trading price of its own spectrum and each buyer subsequently chose one auction to participate. In [38], Wang et al. presented a set of new spectrum double auction mechanisms that were specifically designed for local spectrum markets. Yang et al. in [39] introduced a framework for spectrum double auction mechanisms, which jointly considered spectrum reusability, truthfulness, and profit maximization. In [40], Chen et al. investigated a double auction mechanism for heterogeneous spectrum transaction, where each buyer could bid for specific amount of channels it desired. In [41], a novel truthful double auction mechanism was developed, where the designs of buyers' and sellers' sides were decoupled to capture the different properties of the two sides. Sun et al. in [42] illustrated a coalitional double auction mechanism for spectrum allocation in CR networks, where SUs were partitioned into several coalitions, and the spectrum reusability was executed within each coalition.

2.3.3   Online Spectrum Sharing Mechanisms

Most of the existing spectrum sharing mechanisms are processed in an offline pattern, i.e., the auctioneer collects all the bidding information at the beginning, and then makes the spectrum allocation and payment decisions. Recently, spectrum sharing has also been studied by considering the temporal reusability of spectrum and the uncertain presence of users. Thus, online spectrum sharing mechanism has attracted more and more attention. In an online manner, users may submit bids at any time, while the auctioneer has to make allocation and payment decisions immediately without future information.

In [43], Deek et al. proposed an online spectrum auction framework for CR networks that allocated spectrum efficiently by exploring both spatial and time reusability while resisting buyers from misreporting their valuations and time requirements. Wang et al. in [44] modeled the arrivals of SUs' spectrum requests as Poisson processes and designed a general framework for truthful online double spectrum allocation. Sodagari et al. in [45] investigated a truthful mechanism for expiring spectrum sharing where the property of collusion-resistance was proved in detail. In [46], Xu et al. analyzed a semi-truthful online frequency allocation mechanism where PUs can sublease spectrum to SUs and preempt any existing spectrum usages with some compensation. Li et al. in [47] studied the social welfare maximization problem for serving SUs with various delay tolerances, and compared the performance of the online mechanism with the optimal offline allocation. In [48], an online spectrum auction mechanism with cross-layer decision making and randomized winner determination was proposed, which could achieve truthfulness in expectation and close-to-optimal social welfare in polynomial time complexity.

References

1. N. Nisan, A. Ronen, Algorithmic mechanism design (extended abstract), in *Proceedings of ACM STOC* (1999), pp. 129–140
2. A. Sen, The theory of mechanism design: an overview. Econ. Pol. Wkly. **42**, 8–13 (2007)
3. P.R. Milgrom, *Putting Auction Theory to Work* (Cambridge University Press, Cambridge, 2004)
4. S. Sengupta, M. Chatterjee, An economic framework for dynamic spectrum access and service pricing. IEEE/ACM Trans. Netw. **17**(4), 1200–1213 (2009)
5. Y. Zhang, C. Lee et al., Auction approaches for resource allocation in wireless systems: a survey. IEEE Commun. Surv. Tutorials **15**(3), 1020–1041 (2013)
6. P. Klemperer, What really matters in auction design. J. Econ. Perspect. **16**(1), 169–189 (2002)
7. M.J. Osborne, A. Rubinstein, *A Course in Game Theory* (MIT Press, Cambridge, 1994)
8. V. Krishna, *Auction Theory* (Academic, San Diego, 2009)
9. R.B. Myerson, Incentive compatibility and the bargaining problem. Econometrica **47**, 61–73 (1979)
10. N. Nisan, T. Roughgarden et al., *Algorithmic Game Theory* (Cambridge University Press, Cambridge, 2007)
11. W. Vickrey, Counterspeculation, auctions, and competitive sealed tenders. J. Financ. **16**(1), 8–37 (1961)
12. E.H. Clarke, Multipart pricing of public goods. Public Choice **11**(1), 17–33 (1971)
13. T. Groves, Incentives in teams. Econometrica **41**, 617–631 (1973)
14. A. Pekeč, M.H. Rothkopf, Combinatorial auction design. Manag. Sci. **49**(11), 1485–1503 (2003)
15. N. Nisan, A. Ronen, Computationally feasible vcg mechanisms, in *Proceedings of ACM EC*, vol. 17, no. 20 (2000), pp. 242–252
16. D. Lehmann, L.I. Oćallaghan, Y. Shoham, Truth revelation in approximately efficient combinatorial auctions. J. ACM **49**(5), 577–602 (2002)
17. V. Paulsen, *Completely Bounded Maps and Operator Algebras*, vol. 78 (Cambridge University Press, Cambridge, 2002)
18. J. Huang, R.A. Berry, M.L. Honig, Auction-based spectrum sharing. Mob. Netw. Appl. **11**(3), 405–418 (2006)
19. I. Kash, R. Murty, D. Parkes, Enabling spectrum sharing in secondary market auctions. IEEE Trans. Mob. Comput. **13**(3), 556–568 (2014)
20. L. Gao, J. Huang et al., An integrated contract and auction design for secondary spectrum trading. IEEE J. Sel. Areas Commun. **31**(3), 581–592 (2013)
21. H.-J. Lim, M.-G. Song, G.-H. Im, Cooperation-based dynamic spectrum leasing via multi-winner auction of multiple bands. IEEE Trans. Commun. **61**(4), 1254–1263 (2013)
22. S.-C. Zhan, S.-C. Chang et al., Truthful auction mechanism design for short-interval secondary spectrum access market. IEEE Trans. Wirel. Commun. **13**(3), 1471–1481 (2014)
23. T. Chen, S. Zhong, Truthful auctions for continuous spectrum with variable bandwidths. IEEE Trans. Wirel. Commun. **13**(2), 1116–1128 (2014)
24. O. Naparstek, A. Leshem, Fully distributed optimal channel assignment for open spectrum access. IEEE Trans. Signal Process. **62**(2), 283–294 (2014)
25. Z. Zheng, G. Chen, A strategy-proof combinatorial heterogeneous channel auction framework in noncooperative wireless networks. IEEE Trans. Mobile Comput. **14**(6), 1123–1137 (2015)
26. C. Li, Z. Liu et al., Two dimension spectrum allocation for cognitive radio networks. IEEE Trans. Wirel. Commun. **13**(3), 1410–1423 (2014)
27. F. Wu, Q. Huang et al., Towards privacy preservation in strategy-proof spectrum auction mechanisms for noncooperative wireless networks. IEEE/ACM Trans. Netw. **23**(4), 1271–1285 (2015)
28. F. Rebecchi, M. Dias de Amorim et al., Data offloading techniques in cellular networks: a survey. IEEE Commun. Surv. Tutorials **17**(2), 580–603 (2015)

29. W. Khan, Y. Xiang et al., Mobile phone sensing systems: a survey. IEEE Commun. Surv. Tutorials **15**(1), 402–427 (2013)
30. X. Zhuo, W. Gao et al., An incentive framework for cellular traffic offloading. IEEE Trans. Mob. Comput. **13**(3), 541–555 (2014)
31. W. Dong, S. Rallapalli et al., iDEAL: incentivized dynamic cellular offloading via auctions. IEEE/ACM Trans. Netw. **22**(4), 1271–1284 (2014)
32. I. Koutsopoulos, Optimal incentive-driven design of participatory sensing systems, in *Proceedings of IEEE INFOCOM* (2013), pp. 1402–1410
33. R.B. Myerson, Optimal auction design. Math. Oper. Res. **6**(1), 58–73 (1981)
34. Z. Feng, Y. Zhu et al., TRAC: truthful auction for location-aware collaborative sensing in mobile crowdsourcing, in *Proceedings of IEEE INFOCOM* (2014), pp. 1231–1239
35. X. Zhou, H. Zheng, TRUST: a general framework for truthful double spectrum auctions, in *Proceedings of IEEE INFOCOM* (2009), pp. 999–1007
36. X. Feng, Y. Chen et al., TAHES: a truthful double auction mechanism for heterogeneous spectrums. IEEE Trans. Wirel. Commun. **11**(11), 4038–4047 (2012)
37. L. Gao, Y. Xu, X. Wang, MAP: multiauctioneer progressive auction for dynamic spectrum access. IEEE Trans. Mob. Comput. **10**(8), 1144–1161 (2011)
38. W. Wang, B. Liang, B. Li, Designing truthful spectrum double auctions with local markets. IEEE Trans. Mob. Comput. **13**(1), 75–88 (2014)
39. D. Yang, X. Zhang, G. Xue, Promise: a framework for truthful and profit maximizing spectrum double auctions, in *Proceedings of IEEE INFOCOM* (2014), pp. 109–117
40. Y. Chen, J. Zhang et al., TAMES: a truthful double auction for multi-demand heterogeneous spectrums. IEEE Trans. Parallel Distrib. Syst. **25**(11), 3012–3024 (2014)
41. W. Dong, S. Rallapalli et al., Double auctions for dynamic spectrum allocation, in *Proceedings of IEEE INFOCOM* (2014), pp. 709–717
42. G. Sun, X. Feng et al., Coalitional double auction for spatial spectrum allocation in cognitive radio networks. IEEE Trans. Wirel. Commun. **13**(6), 3196–3206 (2014)
43. L. Deek, X. Zhou et al., To preempt or not: tackling bid and time-based cheating in online spectrum auctions, in *Proceedings of IEEE INFOCOM* (2011), pp. 2219–2227
44. S. Wang, P. Xu et al., Toda: truthful online double auction for spectrum allocation in wireless networks, in *Proceedings of IEEE DySPAN* (2010), pp. 1–10
45. S. Sodagari, A. Attar, S. Bilen, On a truthful mechanism for expiring spectrum sharing in cognitive radio networks. IEEE J. Sel. Areas Commun. **29**(4), 856–865 (2011)
46. P. Xu, X.-Y. Li, TOFU: semi-truthful online frequency allocation mechanism for wireless networks. IEEE/ACM Trans. Netw. **19**(2), 433–446 (2011)
47. S. Li, Z. Zheng et al., Maximizing social welfare in operator-based cognitive radio networks under spectrum uncertainty and sensing inaccuracy, in *Proceedings of IEEE INFOCOM* (2013), pp. 953–961
48. H. Li, C. Wu, Z. Li, Socially-optimal online spectrum auctions for secondary wireless communication, in *Proceedings of IEEE INFOCOM* (2015), pp. 2047–2055

Chapter 3
Recall-Based Spectrum Auction Mechanism

3.1 Introduction

Most of the existing works in dynamic spectrum sharing commonly assumed that the auctioned spectrum resource would be exclusively occupied by the winning spectrum buyers. Such assumption poses a dilemma for the licensed spectrum owners: either auction off unused spectrum bands and get auction revenue at the risk of sudden increases in demand from PUs, or reserve spectrum uneconomically. To address this issue, the idea of *dynamic spectrum recall* has been introduced [1, 2], by which PUs are granted with the highest spectrum access priority so that the auctioned spectrum bands can be recalled from the winning spectrum buyers if necessary.

In this chapter, we present a multi-channel recall-based spectrum auction mechanism for CR networks consisting of one PBS and multiple SUs. Each SU has heterogeneous QoS requirements in terms of spectrum demands and spectrum stability requirements. We begin our discussion with single-winner auction and then extend it to the case with multiple winners. In both scenarios, SUs privately determine their bids based on both the auction information from the PBS and their own preferences including their spectrum demands and stability requirements. For the single-winner spectrum auction, the SPSB mechanism is adopted, whereas in the multi-winner auction, VCG mechanism is applied in the payment design to match the requirements of the formulated combinatorial auction. For both cases, the private valuation of spectrum for each SU is defined, and the optimal strategies for both SUs and the PBS are explored. Both analytical and simulation results show that the recall-based spectrum auction mechanism can improve spectrum utilization and the auction revenue of the PBS, while guaranteeing SUs' heterogeneous QoS requirements.

The rest of this chapter is organized as follows: Sect. 3.2 describes the considered system model and summarizes all important notations used in this chapter.

© The Author(s) 2016
C. Yi, J. Cai, *Market-Driven Spectrum Sharing in Cognitive Radio*,
SpringerBriefs in Electrical and Computer Engineering,
DOI 10.1007/978-3-319-29691-3_3

Section 3.3 defines the single-winner spectrum auction and analyzes the optimal strategies in *recall-based single-winner spectrum auction (RSSA)*. The extension to a multi-winner case, called *recall-based multi-winner spectrum auction (RMSA)*, is introduced in Sect. 3.4. Section 3.5 provides the performance analyses on the auction revenue of the PBS and SUs' utilities in both RSSA and RMSA mechanisms. Numerical results are shown in Sect. 3.6. Finally, a brief summary is presented in Sect. 3.7.

3.2   System Model

Consider a CR network with N SUs that opportunistically access the unused channels of a PBS. The PBS owns total C units of homogenous and indivisible channels. Assume that each PU only requires one channel and the PUs with channel demands will generate a queue at the PBS. We further assume that all PUs obey the first-come-first-serve (FCFS) rule. If all available channels of the PBS have been fully occupied, newly arrived PUs have to wait in the queue. Without loss of generality, the PUs arrive at the PBS following a Poisson process with arrival rate λ so that the interarrival times are independent and identically distributed (i.i.d.) random variables with an exponential distribution. Furthermore, assume that the PUs' channel occupancy times are also i.i.d. exponential random variables with service rate μ. Thus, the channel service of PUs can be viewed as an M/M/m queueing system [3], as shown in Fig. 3.1, where "M" refers to "Markov process" and m denotes the number of channels for PUs. A channel is considered as "idle" if it was not occupied by any PUs; otherwise, it is "busy." Note that SUs have no information about PUs' random activities.

The PBS leases certain number of channels to SUs and, at the same time, provides its PUs with a QoS guarantee. Here, the mean waiting time in the queue is defined as the measurement of the QoS for PUs. Specifically, the mean waiting time of PUs, M_w, cannot be greater than a certain threshold γ. Due to the randomness of PUs' arrivals, if the PBS decides to auction off some unused channels for economic revenue from SUs, it may suffer a risk that there are no enough channels to deal with a sudden increase in PUs' spectrum demands. By considering the higher priority of PUs in CR networks, spectrum recall is allowed for the PBS, i.e., the PBS can recall some channels from the winning SU(s) in order to satisfy its own PUs' demands when necessary. In this way, the newly arrived PUs need to wait if and only if there are no idle channels in the PBS and no more channels can be recalled. Note that, recalled channels will not be returned to SUs until next round of auction. Of course, the auction winner(s) will get corresponding compensation if their channels were recalled by the PBS.

Different from traditional works, in this system, SUs are heterogeneous in spectrum demands and stability requirements. Furthermore, each SU is assumed to work on an integral number of licensed channels. Such assumption is commonly in the literature, such as [4], which considered the application of Microsoft KNOWS

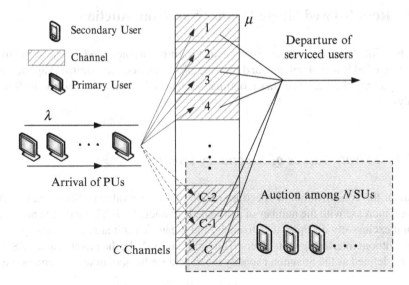

Fig. 3.1 System model of recall-based spectrum auction

Table 3.1 Important notations in this chapter

Notation	Meaning
m	Number of channels needed by the PUs
M_w	Mean waiting time of PUs in the queue
C	Total number of channels owned by the PBS
C_a	Number of auctioned channels
C_r	Maximum number of channels recalled
$C_{r,a}$	Actual number of channels recalled
C_i	Number of channels SU i demands
$C_{r,i}$	Actual number of channels recalled from SU i
ρ_i	Risk factor of SU i in single-winner auction
θ_i	Spectrum stability factor of SU i in multi-winner auction

prototype [5]. Let each SU i have a spectrum demand C_i and a value V_i for C_i channels. Each SU i submits a sealed bid b_i according to its demand to maximize its expected utility.

The auction is carried out frame by frame, and each frame has a length of T. The following discussion is limited to small region networks [6], i.e., all SUs are located within the interference range of each other; hence, no spectrum reuse among SUs within a frame is considered. At the beginning of each T, there is a small period $\tilde{T} \ll T$ used for channel auction.

For convenience, Table 3.1 lists some important notations used in this chapter.

3.3 Recall-Based Single-Winner Spectrum Auction

We first illustrate the auction mechanism for single-winner model. The valuation function of SUs is first defined, and then the SPSB auction mechanism is applied as the payment scheme. After these, optimal strategies for both SUs and the PBS are analyzed.

3.3.1 Private Values of Secondary Users

Without the consideration of spectrum recall, the private value of SU i, i.e., $v_i(C_i)$, should increase with the number of demanded channels C_i. If SU i wins the auction, then it exclusively occupies the channels, and it can transmit at any available power level without interfering with others. Thus, similar to [1, 4], the private value of SU i can be defined as the Shannon capacity it can achieve by obtaining C_i channels, i.e.,

$$v_i(C_i) = C_i \mathcal{T}_w \log_2 \left(1 + \frac{P_t}{n_0 C_i \mathcal{T}_w} \right), \quad \text{for } C_i \geq 0, \tag{3.1}$$

where $\mathcal{T}_w$ is the bandwidth per channel, P_t denotes the unified transmission power of all SUs, and n_0 indicates the spectral density of noise.

Now, let us consider the situation with a recall-based PBS. In this case, the PBS first classifies C channels into two categories, i.e., C_a and $C - C_a$, at the beginning of each auction frame, where C_a channels are auctioned while the remaining $C - C_a$ channels are reserved for its PUs. For the purpose of protecting its own PUs, the PBS also determines C_r, which is the maximum number of channels that can be recalled. In other words, the PBS has at most $C - C_a + C_r$ channels for PUs in the following frame in order to guarantee that the average waiting time of PUs won't be greater than the threshold γ. Obviously, C_r should be less than or equal to C_a.

Apparently, for any SU i, its utility will not decrease with the channel recall if $C_i \leq C_a - C_r$. It means that even under the maximum channel recalls, there is no impact on SU i if it wins the auction. However, if $C_a - C_r < C_i \leq C_a$, the channel recall by the PBS will introduce a reduction on the winning SU i's utility. It is not difficult to find that under the worst case, the maximum number of recalled channels from SU i is equal to $C_i - (C_a - C_r)$. In order to reflect the impacts on SUs due to channel recalls, a particular parameter ρ ($0 \leq \rho \leq 1$), called the risk factor, is introduced. Hence, if $C_a - C_r < C_i$, the benefit of SU i to obtain C_i channels in the recall-based system can be defined as

$$v_i'(C_i) = \left[1 - \frac{C_i - (C_a - C_r)}{C_i} (1 - \rho_i) \right] \times C_i \mathcal{T}_w \log_2 \left(1 + \frac{P_t}{n_0 C_i \mathcal{T}_w} \right), \tag{3.2}$$

where $[C_i - (C_a - C_r)]/C_i$ represents the maximum channel recall ratio on the only winner SU i. Obviously, $v_i'(C_i)$ is a decreasing function of C_r, which matches the intuition that the more channels the PBS declares to recall, the lower private values

SUs may have. The parameter ρ_i is used to reflect different attitudes from SUs toward the potential channel recall. SU with larger ρ is more willing to take risk in this recall-based system and has less concern about the channel recall. Note that ρ is a system factor and cannot be changed by SUs arbitrarily. In fact, such factor heavily depends on the SU's traffic type and its QoS requirements.

Based on the previous discussions, the definition of single-minded SUs in the single-winner spectrum auction with different private value functions can be interpreted as follows.

Definition 3.1. For C homogeneous channels and SU i with valuation V_i, the SU i is single-minded if there is a number of auctioned channels C_a and a number of maximum recalled channels C_r such that

$$V_i = \begin{cases} v_i(C_i), & \text{if} \quad 0 \leq C_i \leq C_a - C_r; \\ v_i'(C_i), & \text{if} \quad C_a - C_r < C_i \leq C_a. \end{cases} \tag{3.3}$$

Note that it is meaningless for SU i to report a demand $C_i > C_a$ since such kind of request can never be satisfied. Hence, all demands from SUs can be assumed to be bounded by C_a. Then, from SU's perspective, there are two distinct outcomes: (1) It gets all channels it demands, i.e., C_i, and does not need to worry about the channel recall; (2) It gets all channels it demands but evaluates C_i with consideration of channel recalls.

3.3.2 Optimal Strategies in RSSA

Since all bids are considered to be sealed, SU i does not know the bids from others. However, it is natural for all SUs and the PBS to know that all bids follow the same valuation function defined in (3.3) except their private information. Further assume that no SU would misreport its channel requests (i.e., C_i). Such assumption is widely used in auctions with heterogeneous bidding requests [7, 8]. If the PBS only allows one winner, no matter how many channels are demanded from each SU, all C_a channels are auctioned to the single winner. Thus, the PBS can simply consider that the auction consists of same-demanded SUs, although each SU may have its specific demand and valuation.

3.3.2.1 Optimal Strategies of SUs

By adopting the SPSB auction mechanism such that the buyer with the highest bid wins the auction but pays the second highest bid, the optimal strategy of each SU is to bid the true valuation of its demanded channels, i.e.,

$$b_i^{single} = V_i. \tag{3.4}$$

Similar to the private value function in (3.3), SU i has two outcomes of bids according to the values of C_i, C_a and C_r. In the first condition, $b_i^{single} = C_i T_w \log_2(1 + P_t/n_0 C_i T_w)$, and the channel recall has no impact on SU i. In this case, the bid is monotonically increased with its spectrum demand C_i. In the second condition, $b_i^{single} = \{1 - (1 - \rho_i)[C_i - (C_a - C_r)]/C_i\} C_i T_w \log_2(1 + P_t/n_0 C_i T_w)$, and the channel recall will affect the service of SU i. In this case, the bid varies with SU's spectrum demand C_i, the maximum recall ratio on SU i and its risk factor ρ_i. The bid would increase as ρ_i increases. That is to say that SU, which is not much concerned on the impact of channel recall so as having a larger ρ, will bid higher in the auction.

3.3.2.2 Auction Information Broadcasting by the PBS

In the RSSA, SU i with highest bidding price wins the auction and pays the second highest bid, i.e., b_{2nd}. Let $C_{r,a}$ and $C_{r,i}$ be the number of channels the PBS recalls after the auction and the actual number of channels recalled from SU i, respectively. Note that for the winning SU i, its demand should be less than the total number of auctioned channels, i.e., $0 < C_i \le C_a$. Thus, we have

$$C_{r,i} = \begin{cases} 0, & \text{if} \quad C_i \le C_a - C_{r,a}; \\ C_i - (C_a - C_{r,a}), & \text{if} \quad C_i > C_a - C_{r,a}. \end{cases} \tag{3.5}$$

Hence, the PBS will compensate $C_{r,i} b_{2nd}/C_i$ back to the winning SU. Then, the final auction revenue of the PBS can be expressed as

$$R_a^{single} = \frac{C_i - C_{r,i}}{C_i} b_{2nd}. \tag{3.6}$$

Since the PUs' QoS will be always protected because of the spectrum recall, we have

$$\max_{C_a} U_{PBS} = \max_{C_a} R_a^{single}, \tag{3.7}$$

where U_{PBS} denotes the utility of the PBS.

Therefore, the optimal strategy for the PBS is to choose the highest bid and maximize its auction revenue. Since C_i is bounded by C_a, and relaxing C_a always produce a nondecreasing R_a^{single}, C_a should be as large as possible, or in other words, the PBS should auction all its idle channels at the beginning of each frame.

In fact, the maximum number of recalled channels C_r can also be determined given $M_w \le \gamma$. For an M/M/m queue with the arrival rate λ and the service rate μ, the minimum number of channels needed by the PUs, i.e., m, can be obtained by

$$M_w = \frac{Q(m, \zeta)}{\mu(m - \zeta)} \le \gamma, \tag{3.8}$$

where $\zeta = \lambda/\mu$, and $\mathcal{Q}(m, \zeta)$ is the queueing probability which can be calculated as

$$\mathcal{Q}(m, \zeta) = \frac{\zeta^m/m!}{[(m - \zeta)/m]\sum_{r=0}^{m-1}(\zeta^r/r!) + \zeta^m/m!}. \tag{3.9}$$

Suppose $C \geq m$. Thus, in order to guarantee the QoS of PUs, the minimum number of total reserved channels at the PBS should satisfy the condition that $m \leq (C - C_a) + C_r$ or $C_r \geq m - (C - C_a)$. In addition, C_r should not be greater than the number of auctioned channels. Therefore, C_r should be ranged as $C_a \geq C_r \geq m - (C - C_a)$. Since the derivation of m has guaranteed the QoS of PUs, the PBS has no intention to reserve more channels, i.e., $C_r = m - (C - C_a)$.

However, the PBS may cheat in the auction if its utility can be further improved. According to the discussions above, the PBS may benefit by misreporting the amount of auctioned channels $C_a' > C_a$ or misreporting the maximum amount of recalled channels $C_r' < C_r$. Intuitively, the latter kind of cheating is much more harmful and possible to happen in practice, since reporting $C_a' > C_a$ will be immediately realized by the winning SU at the beginning of each auction frame, while misreporting $C_r' < C_r$ can only be discovered at the end of each frame.

3.3.2.3   Amendments of SUs' Private Values

As mentioned above that the PBS may use mendacious information (i.e., reporting an untruthful C_r) for more revenue. Specifically, given a constant C_a, a smaller C_r declared by the PBS indicates more available spectrum for SUs. As a result, SUs will bid higher values. However, if the amount of recalled channels is larger than C_r at the end of the auction, SUs will suffer a loss. They will notice that the PBS cheat them by broadcasting untruthful auction information in order to gain more auction revenue.

For protecting SUs in repeated auctions, each SU i is allowed to add a belief index $\varphi_i \leq 1$ to its valuation function which denotes its belief on the truthfulness of the PBS. A small value of φ_i implies that SU i lacks trust on the PBS. With φ_i, the private value function of SU i can be modified as

$$V_i' = \varphi_i \cdot V_i. \tag{3.10}$$

At the end of the $(\ell - 1)$th auction, SU i updates its belief index from $\varphi_i(\ell - 1)$ to $\varphi_i(\ell)$ for the coming ℓth auction according to Algorithm 1. In repeated games, trigger strategy is widely adopted for punishing possible strategies deviations [9]. In this problem, the update rule of φ can be designed as the SUs' trigger strategy for punishing the PBS's untruthful behaviors.

In Algorithm 1, $C_{r,a}(\ell)$ is the practical amount of channels recalled in ℓth round of auction and $\varphi(\ell)$ is the SU's belief index in the ℓth round. $\varphi_0 = 1$ is the initial value of belief index which means that SU i trusts the PBS. $\alpha < 1$ is set to be a discount ratio. SUs will decrease their belief index in the next round if they notice

---

Algorithm 1 Update rule of belief index

1: **if** $C_{r,a}(\ell) > C_r(\ell)$ **then**
2:    $\varphi(\ell + 1) = \alpha\varphi(\ell)$;
3: **else if** $C_{r,a}(\ell - k) < C_r(\ell - k), k = 0, 1, \ldots, \Delta\ell - 1$ **then**
4:    $\varphi(\ell + 1) = \varphi_0$;
5: **else**
6:    $\varphi(\ell + 1) = \varphi(\ell)$;
7: **end if**

---

that they have been cheated. Moreover, this reduction will be kept for $\Delta\ell$ rounds after the PBS's rehabilitation (i.e., recall less channels than C_r). In other words, one time deception causes $\Delta\ell$ times punishment.

3.3.3 Time-Line of RSSA

The detailed time-line of the RSSA is listed as follows:

- At the beginning of each frame, the PBS broadcasts the auction information including the number of auctioned channels C_a and the maximal recalls C_r based on its current service state. The settings of C_a and C_r can be found in Sect. 3.3.2.
- Each SU i receives the auction information and sets up a value V_i based on its own spectrum demand C_i, risk factor ρ_i and the maximum channel recall ratio on itself, i.e., $[C_i - (C_a - C_r)]/C_i$. Then, SUs submit sealed bids and their specific demands to the PBS.
- The PBS determines the only winner by selecting the SU with highest bidding price and charges it with the second highest bid b_{2nd}.
- After the auction, the PBS can recall channels from the winning SU i if necessary to satisfy its own sudden increase in spectrum demand. At the end of T, the PBS refunds SU i with $C_{r,i}b_{2nd}/C_i$.

3.4 Recall-Based Multi-Winner Spectrum Auction

Since the demand of each SU i, i.e., C_i, is independent of the number of auctioned channels C_a, it is very likely that the auctioned channels C_a cannot be fully utilized by one winning SU. Thus, the auction revenue can be enhanced if the PBS picks more than a single winner. However, the allowance of multiple winners makes the spectrum auction become a more complicated combinatorial auction problem.

In the payment design of the RMSA, VCG mechanism is adopted. Although VCG mechanism cannot guarantee the maximum auction revenue for the PBS [10], it is the basic payment mechanism in combinatorial auction that can ensure efficiency, incentive compatibility and individual rationality. Note that revenue-maximizing combinatorial auction mechanism or any approximate mechanism, such as virtual valuation combinatorial auctions (VVCA) [11] or LOS mechanism [12], can also be applied in the RMSA.

3.4.1 Strategies of Secondary Users in RMSA

Without channel recall, the private value on channel demand C_i of SU i in multi-winner auction is the same as that in (3.1), i.e., $v_i(C_i) = C_i T_w \log_2 (1 + P_t/n_0 C_i T_w)$ for $C_i \geq 0$.

With channel recall, the PBS needs to announce C_a and C_r at the beginning of each frame. Different from the single-winner case where each SU can figure out the maximum number of channels recalled from itself if it won the auction, such information is not available in multi-winner auction because the number of channels recalled from a winning SU is not only determined by its own demand, but also by the demands of other winners.

Let $W \subseteq \{1, 2, \ldots, N\}$ be the set of winners. Different from single-winner case, since the auctioned channels C_a may not be fully utilized by winners in W, the maximum channel recall ratio on W equals

$$\gamma^{multiple} = \frac{C_r - (C_a - \sum_{i \in W} C_i)}{C_a}. \tag{3.11}$$

Same as in the RSSA, each SU needs to evaluate its private value toward its spectrum demand based on $\gamma^{multiple}$. However, the term $(C_a - \sum_{i \in W} C_i)$ is unpredictable since W cannot be determined before the auction. Thus, $\gamma^{multiple}$ is approximated as C_r/C_a.

Similar to the risk factor in the RSSA, let us define $\theta_i \in [0, 1]$ as SU i's spectrum stability factor in its private value function. Although θ_i also reflects the attitude of SU i toward channel recall, the physical meaning of θ in the RMSA is different from ρ in the RSSA. In the single-winner case, since the maximum channel recall ratio on the single winner can be determined before auction, the spectrum stability is only determined by the activity of PUs. However, in the multi-winner case, since the maximum channel recall ratio can only be determined at the system level, i.e., C_r/C_a, rather than for each winner, the spectrum stability factor may affect both the winner determination and the channel recall ratio on each winner.

The definition of single-minded SUs in the RMSA is given in the following.

Definition 3.2. For C homogeneous channels and SU i with valuation V_i, SU i is single-minded if there exist a number of auctioned channels C_a and a number of maximum recall C_r such that

$$V_i = v_i''(C_i) = \left[1 - \frac{C_r}{C_a}(1 - \theta_i)\right] \times v_i(C_i), \quad \text{if } 0 \leq C_i \leq C_a. \tag{3.12}$$

Thus, from SU i's perspective, it gets channels it demands, but multiplies a channel stability ratio to its valuation. Similar to the RSSA, the larger C_r the PBS declares, the lower private values SUs may have. Moreover, SUs with different spectrum demands and stability factors will also lead to different private values. Larger θ_i indicates that SU i can be provided a more stable service so as to gain higher utility. Note again that θ is also predetermined by the system based on SUs' traffic types and transmission requirements, and thus cannot be changed arbitrarily by SUs.

In the payment design, the use of VCG mechanism requires that all buyers only know their own private values for their demands and each of them has a quasi-linear utility function.

Proposition 3.1. *For SU $i \in \{1, 2, \ldots, N\}$ with particular spectrum demand C_i and spectrum stability requirement factor θ_i, $U_i = V_i - p_i$ is a quasi-linear utility function, where p_i denotes the payment of SU i in the auction.*

Proof. In order to prove that $u_i = V_i - p_i$ is a quasi-linear utility function, we only need to show that V_i is a concave function of channel demand C_i [13]. Recall that $V_i = [1 - (C_r/C_a)(1 - \theta_i)]C_i T_w \log_2(1 + P_t/(n_0 C_i T_w))$, if $0 \leq C_i \leq C_a$. Since the ratio caused by channel recall, $1 - (C_r/C_a)(1 - \theta_i)$, is independent of C_i, the concavity and convexity of V_i only depends on the formula of Shannon capacity. In fact, it can be directly proved that the capacity $C_i T_w \log_2(1 + P_t/(n_0 C_i T_w))$ is an increasing, concave function of bandwidth $C_i T_w$ [14]. Thus, V_i is a concave function of C_i, and thus u_i is a quasi-linear utility.   □

According to VCG mechanism, truthful bidding maximizes any buyer's utility regardless of other buyer' strategies. Hence, all the SUs will truthfully bid in the multi-winner spectrum auction by honestly telling the PBS their private values, i.e.,

$$b_i^{multiple} = V_i. \tag{3.13}$$

3.4.2 Actions of the Primary Spectrum Owner

Similar to the analysis in Sect. 3.3.2 for determining C_a and C_r in the RSSA mechanism, the PBS will auction all its idle channels and announce a maximum recall quantity C_r based on PUs' QoS requirement. For simplicity, let us assume that the PBS in the RMSA is always truthful (otherwise the trigger amendment on SU's utility as shown in Sect. 3.3.2.3 can also be applied here). Here, we focus on the winner determination of the combinatorial auction and the payment charged from each winner. Furthermore, a new channel recall scheme is introduced to achieve some level of fairness in spectrum sharing among heterogeneous SUs.

3.4.2.1 Winner Determination and Payment Design

In the RMSA, each SU reports the PBS its sealed bid and specific spectrum demand. The PBS determines the winners by solving the following optimization problem.

Given bids $B = \{b_1, b_2, \ldots, b_N\}$ and spectrum demands $\{C_1, C_2, \ldots, C_N\}$ from all SUs, the PBS determines the winners such that

$$\max_{\{x_i\}, \forall i \in N} S_B^C = \sum_{i=1}^N b_i x_i$$

$$s.t. \quad \sum_{i=1}^N C_i x_i \leq C_a, \tag{3.14}$$

where

$$x_i = \begin{cases} 1, & \text{if SU } i \text{ is the winner of the auction;} \\ 0, & \text{otherwise.} \end{cases}$$

The optimization problem (3.14) aims to find the set of winners $W = \{i | x_i = 1, \forall i \in N\}$ such that the sum of their bids received by the PBS is maximized under the constraint that their total spectrum demand is less than or equal to the number of auctioned channels C_a. Furthermore, since we have assumed that SUs are single-minded so that SU i can either get all spectrum it demands or nothing, problem (3.14) is actually a $0 - 1$ single knapsack problem that can be solved to optimality in pseudo-polynomial time by using dynamic programming or branch and bound algorithm [15]. Note that the availability of optimal solution to (3.14) guarantees the feasibility of VCG mechanism.

After deciding the set of winners, the PBS charges the winning SUs according to the VCG mechanism. The payment of SU i is

$$p_i = S_{B \backslash \{b_i\}}^C - S_{B \backslash \{b_i\}}^{C \backslash \{C_i\}}, \tag{3.15}$$

where $S_{B \backslash \{b_i\}}^C$ denotes the maximum welfare if SU i does not participate in the auction and $S_{B \backslash \{b_i\}}^{C \backslash \{C_i\}}$ denotes the maximum welfare if SU i does not participate and it takes out its demanded C_i channels from the total C channels in the auction. The details of payment rule in VCG mechanism can be found in Sect. 2.2.4.

3.4.2.2 Channel Recall Scheme

The VCG mechanism is actually designed for buyers with fixed private values. However, in the considered system model, the utilities of winning SUs may decrease after the auction because of channel recalls. Hence, a careful design of the spectrum recall scheme is required for the RMSA.

For explanation purpose, we first introduce a simple definition of fairness index.

Definition 3.3 (Min-Max Fairness). For each winner $i \in W$, if the actual number of channels recalled on SU i is less than its spectrum demand, i.e. $C_{r,i} < C_i$, a resource allocation index can be defined as

$$f_i = \frac{C_i - C_{r,i}}{p_i}, \qquad (3.16)$$

where $C_i - C_{r,i}$ indicates the actual number of channels SU i obtained and p_i is the payment. Given f_i, $i \in W$, a min-max fairness index can be defined as

$$\mathcal{I}_{min-max} = \frac{min\{f_i\}}{max\{f_i\}}, \quad \forall i \in \{i | i \in W, C_{r,i} < C_i\}. \qquad (3.17)$$

Obviously, according to this definition, the spectrum allocation is more fair when $\mathcal{I}_{min-max}$ tends to 1.

Proposition 3.2. *The VCG mechanism is unfair under the situation that multiple homogeneous channels are auctioned among SUs with different stability require-ments and a same recall ratio $C_{r,a}/C_a$ on multiple winners is applied.*

Proof. Consider the system with only two winners, SUs i, j, both of which have same spectrum demands, i.e., $C_i = C_j$. According to (3.12), the difference of their private values only depends on the spectrum stability factor θ. Assume that $\theta_i > \theta_j$. Then, SU i has a larger private value than SU j, which leads to a larger bid, i.e, $b_i > b_j$. With the VCG mechanism of item allocation and payment design, it is easy to verify that SU i and SU j will get the same number of channels, but with $p_i > p_j$. Since the recall ratio is the same on both SUs i and j, i.e., $C_{r,a}/C_a$, the fairness index can be calculated as

$$\mathcal{I} = \frac{f_i}{f_j} = \frac{C_i(1 - C_{r,a}/C_a)}{p_i} \times \frac{p_j}{C_j(1 - C_{r,a}/C_a)} = \frac{p_j}{p_i}. \qquad (3.18)$$

Apparently, this scheme is not fair, especially for the case that $p_i \gg p_j$ when $\theta_i \gg \theta_j$.   □

This proposition indicates that applying same recall ratio on multiple winners is not reasonable for SUs with different spectrum stability factors.

In addition, the channel recall may also affect the auction revenue of the PBS. According to the VCG mechanism, channel recalls will be evenly distributed among winning SUs. The recall compensation is equal to the product of actual spectrum recall ratio and the sum of payments gained from winners, i.e., $C_{r,a}/C_a \times \sum_{i \in W} p_i$. Thus, the revenue of the PBS can be written as

$$R_{a1}^{multiple} = \left(1 - \frac{C_{r,a}}{C_a}\right) \sum_{i \in W} p_i. \qquad (3.19)$$

Intuitively, the PBS can get more profit and reduce the compensation by recalling more channels from the winners with low payments.

From the above analyses, a simple but effective *pricing-based channel recall scheme* can be applied, which is interpreted as follows.

Assuming that during $t \in [\Delta T, T]$, the winning SU $i \in W$ uses C_i channels, and totally $\sum_{i \in W} C_i$ channels are used by SUs. The PBS can recall channels one by one when necessary. Since the PBS knows the payment of each SU and the details of auction mechanism, it can figure out the unit payoff of each channel. Thus, the channel with lower payoff will be granted with higher priority to be recalled, and the unused channels will be recalled in the first place. At the end of T, the PBS refunds winning SU i with $p_i \times C_{r,i}/C_i$, where $C_{r,i}$ denotes the number of channels which are actually recalled from SU i. Note that $C_{r,i}$ is heterogeneous for each winner, and it is likely that $C_{r,i} = 0$ for the winner with high unit payment for each channel, whereas the channels may be completely recalled for the winner with low unit payment.

3.4.3 Time-Line of RMSA

The detailed time-line of the RMSA can be summarized in the following.

- The PBS broadcasts the auction information including C_a and C_r at the beginning of each frame.
- Each SU i receives the auction information and sets up a value V_i based on its own spectrum demand C_i, stability factor θ_i and the channel recall ratio C_r/C_a. Then, SUs submit sealed bids and their specific spectrum demands to the PBS.
- The PBS determines the winner by solving the optimization problem in (3.14) and charges the winning SU $i \in W$ based on the VCG payment rule in (3.15).
- After the auction, the PBS can recall channels one by one to meet its own sudden increase in spectrum demand. The channel recall follows the scheme illustrated in Sect. 3.4.2.2. At the end of T, the PBS refunds each winning SU i with $p_i \times C_{r,i}/C_i$.

3.5 Performance Analyses

In this section, economic properties of the RSSA and RMSA mechanisms are analyzed in terms of the PBS's auction revenue and SUs' utilities.

3.5.1 Revenue of the Primary Base Station

Since channel recall is enabled at the PBS, the PUs' service will be completely protected. Therefore, the PBS takes no risk on the utility degradation but only benefits from the dynamic spectrum auction. Hence, we can focus on analyzing the auction revenue of the PBS only.

In the considered system model, the arrival of PUs follows Poisson process and spectrum auction is carried out by the PBS frame by frame. We use $u(\ell)$ to represent the number of PUs who are in service at the end of the ℓth frame. Obviously, $u(\ell)$ also indicates the number of busy channels at the beginning of frame $\ell+1$. Thus, the number of auctioned channels at the beginning of the ℓth frame is $C_a = C-u(\ell-1)$. In addition, the actual number of channels recalled during the ℓth frame is $C_{r,a} = u(\ell) - u(\ell - 1) + d(\ell)$, where $d(\ell)$ denotes the number of all departures during that period. Note that only $u(\ell - 1)$ is known by the PBS at the beginning of the ℓth frame, while $u(\ell)$ and $d(\ell)$ are unknown.

For single-winner auction with channel recall, the winner determination will be optimal only if the winner i^* satisfies:

$$i^* = \arg \max_i \frac{C_i - C_{r,i}}{C_i} \times b_i',$$
(3.20)

where

$$b_i' = v_i'(C_i, C_{r,i}) = \left[1 - \frac{C_i - (C_a - C_{r,i})}{C_i}(1 - \rho_i)\right] v_i(C_i).$$
(3.21)

Note that (3.21) is formulated based on the assumption that the accurate amount of channel recall, i.e., $C_{r,i}$, is known at the beginning of the auction. Obviously, the bidding pattern and winner determination in the RSSA may be suboptimal compared to the above case with complete information, since $C_{r,i}$ is actually unknown at the beginning of the RSSA with unknown $C_{r,a}$. Such deficit on the auction revenue for recall-based systems will be presented numerically by the simulation in Sect. 3.6.

Similarly, such problem also exists in multi-winner auction under the VCG mechanism. In fact, the optimal winner determination and spectrum allocation should satisfy the following conditions: Given bids $B = \{b_1'', \ldots, b_i'', \ldots, b_N''\}$, where $b_i'' = v_i(C_i)$, and spectrum demand $\{C_1, C_2, \ldots, C_N\}$, the winner set $W^* = \{i|x_i = 1, \forall i \in N\}$ is determined by

$$\max_{\{x_i\}, \forall i \in N} S_B^C = \sum_{i=1}^N b_i'' x_i$$
(3.22)

$$s.t. \quad \sum_{i=1}^N C_i x_i \leq C_a - C_{r,a},$$

$$x_i = 0/1, \quad \forall i \in \{1, 2, \ldots, N\}.$$

For the same reason that $C_{r,a}$ is unknown at the beginning of the RMSA, it is impossible for the PBS to find the optimal decision, and SUs will not bid non-recall valuations.

We now analyze the performance of the pricing-based channel recall scheme in terms of the PBS's auction revenue. After receiving payments from the winners in W, the PBS rearrange the payments according to an increasing order of the unit price per channel. Let the payment set as $\{p^1, p^2, \ldots, p^{|W|}\}$, where $|W|$ is the number of elements in W, and C_{p^i} be the demand of SU who paid p^i. If SU j^* who paid p^{j^*} is the last one in W whose channel will be completely recalled, j^* can be found as

$$j^* = \arg\min_j \left(\sum_{k=1}^{j} p^k + \frac{C_{r,a} - \sum_{k=1}^{j} C_{p^k}}{C_{p^j+1}} \times p^{j+1} \right). \tag{3.23}$$

Therefore, the PBS's auction revenue under the pricing-based recall scheme can be expressed as

$$R_{a2}^{multiple} = \sum_{i \in W} p_i - \left(\sum_{k=1}^{j^*} p^k + \frac{C_{r,a} - \sum_{k=1}^{j^*} C_{p^k}}{C_{p^{j^*}+1}} p^{j^*+1} \right). \tag{3.24}$$

We can then evaluate the performance by comparing $R_{a1}^{multiple}$ in (3.19) and $R_{a2}^{multiple}$ in (3.24) as

$$\begin{aligned}
\Re &= R_{a2}^{multiple} - R_{a1}^{multiple} \\
&= \sum_{i \in W} p_i - \left(\sum_{k=1}^{j^*} p^k + \frac{C_{r,a} - \sum_{k=1}^{j^*} C_{p^k}}{C_{p^{j^*}+1}} p^{j^*+1} \right) - \left(1 - \frac{C_{r,a}}{C_a} \right) \sum_{i \in W} p_i \\
&= -\left(\sum_{k=1}^{j^*} p^k + \frac{C_{r,a} - \sum_{k=1}^{j^*} C_{p^k}}{C_{p^{j^*}+1}} p^{j^*+1} \right) + \frac{C_{r,a}}{C_a} \sum_{i \in W} p_i \\
&= -\left(\sum_{k=1}^{j^*} p^k + \frac{C_{r,a} - \sum_{k=1}^{j^*} C_{p^k}}{C_{p^{j^*}+1}} p^{j^*+1} \right) + \frac{C_{r,a}}{C_a} \sum_{k=1}^{m} p^k.
\end{aligned} \tag{3.25}$$

Let $\delta_1 = C_{r,a}/C_a \sum_{k=1}^{m} p^k$ indicate the compensation in the auction with evenly distributed channel recall scheme, and $\delta_2 = \sum_{k=1}^{j^*} p^k + p^{j^*+1}(C_{r,a} - \sum_{k=1}^{j^*} C_{p^k})/C_{p^{j^*}+1}$ represent the compensation by the pricing-based channel recall scheme. With the definition in (3.23), we have $\Re = \delta_1 - \delta_2 \geq 0$. Therefore, the pricing-based channel recall scheme outperforms the evenly distributed scheme in increasing the auction revenue of the PBS.

3.5.2 Utilities of Secondary Users

Each SU's utility equals the difference between its gain and payment. At the beginning of the auction, SU i evaluates C_i channels based on the auction information. However, winning SU i may not obtain C_i channels due to the potential channel recall. Therefore, we need to investigate the relation between SU's utility and its private information. Moreover, in order to satisfy the heterogeneous requirements of SUs and provide them a fair spectrum allocation in multi-winner auction, we need to prove that any SU with a higher spectrum stability factor, which results in a higher bid for a unit of spectrum can be guaranteed with a more stable service by the PBS.

3.5.2.1 SU's Utility in RSSA

Consider the case without channel recall first. With the payment rule of SPSB mechanism, the buyer with the highest bid wins, but the price paid is the second highest bid. Thus, the expected utility of SU i is

$$U_i = (V_i - b_{2nd})\text{Pr.}\left\{b_i > \max_{j \neq i} b_j\right\} \tag{3.26}$$

where $V_i - b_{2nd}$ is its net utility and $\text{Pr.}\{b_i > \max_{j \neq i} b_j\}$ is its winning probability. For the system with channel recall, two cases need to be discussed.

- Case 1: $0 \leq C_i \leq C_a - C_{r,a}$. The actual gain of SU i is same as (3.1), i.e., $G_i = v_i(C_i)$.
- Case 2: $C_a - C_{r,a} < C_i \leq C_a$. The actual gain G_i' can be obtained by (3.2), except that the amount of obtained channels C_i is replaced by $C_i - C_{r,i}$, i.e.,

$$G_i' = \left[1 - \frac{C_i - (C_a - C_r)}{C_i}(1 - \rho_i)\right](C_i - C_{r,i})B \log_2\left(\frac{P_t}{n_0(C_i - C_{r,i})B}\right). \tag{3.27}$$

Then, we can derive the expected utility for SU i in the recall-based system as

$$U_i^{single} = \begin{cases} (G_i - b_{2nd})\text{Pr.}\{b_i > \max_{j \neq i} b_j\}, & (3.28) \\[2ex] (G_i' - (1 - \frac{C_{r,i}}{C_i})b_{2nd})\text{Pr.}\{b_i > \max_{j \neq i} b_j\}, & (3.29) \end{cases}$$

with (3.28) and (3.29) corresponding to cases 1 and 2, respectively.

Lemma 3.1. *In the RSSA mechanism, the utility of SU i is not monotonically increased with its spectrum demand C_i.*

Proof. Apparently, U_i^{single} is monotonically increased with C_i in case 1. However, this property would not be maintained when C_i continues to increase. Since SU_i in case 1 can fully utilize C_i channels, but SU i in case 2 is affected by the channel recall, U_i^{single} has a sudden decrease when C_i reaches the threshold $C_a - C_{r,a}$. Therefore, the utility of SU i cannot consecutively increase with C_i from 0 to C_a. $\qquad \square$

The following Lemma shows the impact on SUs' utilities caused by different risk factors.

Lemma 3.2. *In the RSSA mechanism, the SU with a larger value of risk factor ρ has a better service (i.e., higher utility) than the SU with a smaller one.*

Proof. Obviously, the gain and the bid of SU i in both case 1 and case 2 increase with ρ_i, i.e., $\partial G_i/\partial \rho_i > 0$ and $\partial G_i'/\partial \rho_i > 0$, and ρ_i has nothing to do with the compensation. Thus, we have

$$\frac{\partial U_i^{single}}{\partial \rho_i} \geq 0. \tag{3.30}$$

Moreover, $\text{Pr.}\{b_i > \max_{j \neq i} b_j\}$ would also be enhanced when ρ_i is larger. Thus, the SU with larger risk factor ρ has a higher chance to win the auction. $\quad\square$

With above lemmas, the advantages of the RSSA mechanism can be concluded in the following theorem.

Theorem 3.1. *The RSSA mechanism can provide economic incentives for all the SUs to participate in the auction since their utilities are always non-negative and their heterogeneous requirements can be satisfied when they win the competition.*

Proof. Since the RSSA mechanism is originated from the SPSB mechanism, the incentive compatibility and individual rationality are maintained. Moreover, all SUs are assumed to truthfully report their channel demands. With the help of Lemmas 3.1 and 3.2, we can observe that the utility of an SU is strictly related to how much it concerns for channel recall but not its channel demand. Hence, all the SUs will follow the rules in the RSSA. $\quad\square$

3.5.2.2 SU's Utility in RMSA

Similarly, without the consideration of channel recall, the utility of SU i in the RMSA can be expressed as

$$U_i' = (V_i - p_i)\text{Pr.}\{i \in W\}, \tag{3.31}$$

where $V_i - p_i$ is its net utility and $\text{Pr.}\{i \in W\}$ is its winning probability in the knapsack problem of (3.14).

With channel recall, according to the valuation function in (3.12), the gain of winning SU i is

$$G_i'' = \left[1 - \frac{C_r}{C_a}(1 - \theta_i)\right](C_i - C_{r,i})B\log_2\left(\frac{P_t}{n_0 C_i T_w}\right). \tag{3.32}$$

Further considering the compensation of channel recall, the expected utility of SU i in the RMSA can be rewritten as

$$U_i^{multiple} = \left[G_i'' - \left(1 - \frac{C_{r,i}}{C_i}\right) p_i \right] \Pr.\{i \in W\}. \tag{3.33}$$

Lemma 3.3. *In the RMSA mechanism, the utility of SU i can only be ameliorated with the increase in stability factor θ_i.*

Proof. Obviously, the utility of SU i is not monotone with C_i since the number of recalls on SU i, $C_{r,i}$, also increases with the demand C_i. However, $C_{r,i}$ will decrease with the increase in θ_i because of the pricing-based channel recall scheme. That means $\frac{\partial C_{r,i}}{\partial \theta_i} < 0$. Thus,

$$\frac{\partial G_i''}{\partial \theta_i} = \left[1 - \frac{C_r}{C_a}(1 - \theta_i) - \frac{\partial C_{r,i}}{\partial \theta_i} + \frac{C_r}{C_a}(C_i - C_{r,i}) \right] B \log_2(1 + \frac{P_t}{n_0(C_i - C_{r,i})B}) > 0.$$

Although the compensation $C_{r,i}/C_i \cdot p_i$ is monotonically decreased with θ_i, such decrease is less than the increase in G_i'' since the payment is always less than the gain to ensure non-negative utility. Hence, we have

$$\frac{\partial U_i^{multiple}}{\partial \theta_i} \geq 0. \tag{3.34}$$

Moreover, the increase in θ_i will also result in the enhancement of $\Pr.\{i \in W\}$. That means SU i with larger θ_i has higher probability to win the auction and the quantity of recall, $C_{r,i}$, will decrease. In other words, the spectrum occupied by SU i with larger θ_i is more stable.   □

Theorem 3.2. *The RMSA mechanism can provide economic incentives for all the SUs to participate in the auction since their utilities are non-negative and the mechanism also ensures that the heterogeneous requirements of SUs can be satisfied.*

Proof. Since the RMSA mechanism is originated from the VCG mechanism, the incentive compatibility and individual rationality are guaranteed automatically. Moreover, according to Lemma 3.3, the SU with a larger stability factor is granted with a more stable spectrum environment. Thus, the RMSA mechanism can meet the heterogeneous requirements of SUs.   □

3.6 Numerical Results

Here, some simulations are conducted to numerically evaluate the performance of RSSA and RMSA mechanisms.

Consider a CR network with one PBS and N heterogeneous SUs. PUs' arrival rate $\lambda = 2$ and channel service rate $\mu = 0.1$. The threshold γ is set as 6.25×10^{-4} s, and the PBS owns $C = 36$ channels to satisfy the inequality (3.8). The length of each frame $T = 6$ s; hence, the average number of PUs arrive in 1 min is 20 and the mean time of service for each PU is 60 s. These settings are commonly used in the design of the mobile base station [1]. Furthermore, $\mathcal{T}_w = 10^5$ Hz, $n_0 = 2 \times 10^{-10}$ W/Hz and $P_t = 0.01$ W. Note that the number of SUs N, spectrum demands C_i and factors ρ_i, θ_i for each SU i are varied according to the evaluation scenarios.

Figure 3.2 shows the PBS's state information (i.e., the number of "idle" and "busy" channels) at each frame. For each frame, the number of active PUs can be determined by the parameters of queueing system. Since the PBS auctions all the idle channels, the number of auctioned channels is increased when the number of PUs decreases. Moreover, the increase in recall also leads to a decrease in the number of auctioned channels. Since the PBS will be truthful in the long-term auction, it is shown that the announced number of maximum recall is always larger than the number of actual recalls. All the rest simulation results are based on the state information shown in this figure.

In Fig. 3.3, the auction revenue of the PBS is compared between the optimal winner determination as described in (3.20), and the RSSA mechanism. Intuitively, a small-scale network has a higher probability of coincidence that the optimal determination is the same as the decision made by the RSSA mechanism. Therefore, a relatively large network with $N = 50$ SUs is considered in this simulation. Moreover, the demand of each SU is selected randomly from integers 0 to 15 and risk factor is chosen randomly in $[0, 1]$. Figure 3.3 shows that the curve of the PBS's

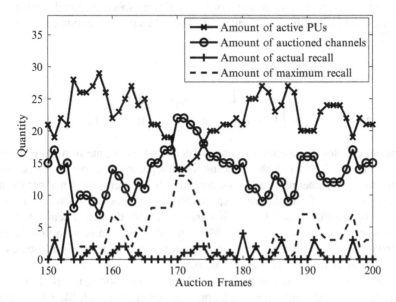

Fig. 3.2 State information of the PBS in different auction frames

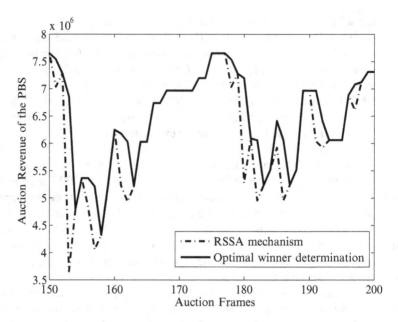

Fig. 3.3 The auction revenue of the PBS in RSSA ($N = 50$)

auction revenue obtained by the RSSA mechanism is still close to the one with optimal winner determination. It indicates that the recall-based system can achieve close to optimal performance for the PBS.

In order to demonstrate the superiority of enabling spectrum recall, we compare the utilities of PBS with and without recall. The PBS's utility function can be defined as [16]

$$U_{PBS} = R_a + R_s - U_{punish} \tag{3.35}$$

$$= \begin{cases} R_a + \omega_s C_s - \omega_p \frac{C_v - C_s}{C_s}, & \text{if } C_v \geq C_s, \\ R_a + \omega_s C_v - \omega_p \frac{C_s - C_v}{C_s}, & \text{if } C_v < C_s, \end{cases}$$

where R_a denotes the auction revenue, R_s denotes the revenue from its own PUs' service, and U_{punish} is a punishment term, which represents the loss due to excessive or insufficient channel reservation. ω_s and ω_p indicate average revenue per PU and the weight index of punishment, respectively. C_v denotes the amount of channel reserved by the PBS before the auction and C_s denotes the actual demand of PUs. In the simulation, we set $\omega_s = \omega_p = 10^6$. In Fig. 3.4, it can be seen that the PBS has lower and more fluctuating utility without recall, which clearly illustrates the improvement by using recall-based system.

Figure 3.5 exhibits the comparison of channel utilization ratio between single-winner auction and multi-winner auction. The demand of each SU ($N = 10$) is selected randomly from integer 1 to 10, and both ρ and θ are chosen randomly

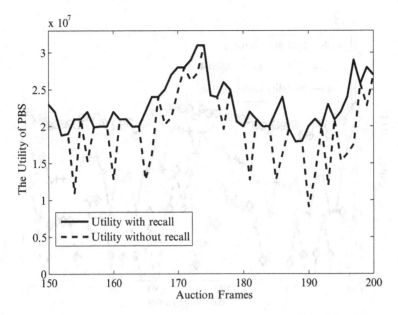

Fig. 3.4 The utility of the PBS in recall-based system ($N = 50$)

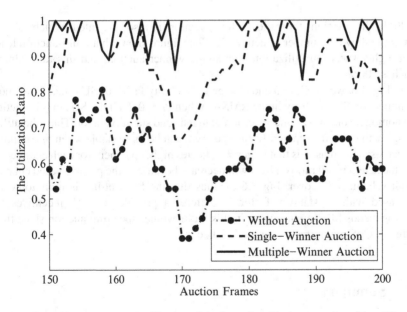

Fig. 3.5 Comparison on spectrum utilization of single- and multi-winner auctions ($N = 10$)

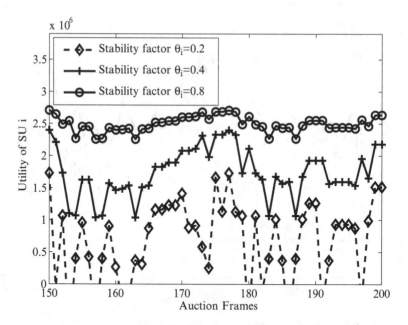

Fig. 3.6 The utility of SU i with different stability factors in RMSA

in $[0, 1]$. In addition, the ratio without auction is also presented, which is only determined by the number of active PUs. The figure shows that multi-auction leads to a higher spectrum utilization than single-winner auction, and this ratio almost reaches 100 %.

In Fig. 3.6, we fix $C_i = 5$ and change the stability factor of SU i, i.e., θ_i, to show its impact on SU i's utility in the RMSA. When $\theta_i = 0.2$, all the SUs in this auction are homogeneous with same spectrum demands and stability factor. Thus, the utility is highly fluctuated. Apparently, the spectrum can be more stable when θ_i continues to increase. The reason is that larger θ_i indicates higher payment for each channel, so that the actual recall ratio on SU i will decrease because of the pricing-based channel recall scheme. Moreover, Fig. 3.6 justifies that the SU's utility is monotonically increased with its stability factor. Therefore, it provides enough incentives for heterogeneous SUs to participate in this multi-winner spectrum auction since their different QoS requirements can be satisfied.

3.7   Summary

In this chapter, a recall-based spectrum sharing in CR networks with a single PBS and multiple heterogeneous SUs has been discussed. Both single- and multi-winner cases were studied. In order to meet the SUs' requirements on spectrum demands and stability, the private value function for each single-minded SU was redefined, and RSSA and RMSA mechanisms were introduced for single- and multi-winner auctions, respectively. Theoretical and simulation results showed that the recall-based auction mechanisms could increase the auction revenue of the PBS, and enhance the spectrum utilization efficiency. Moreover, the heterogeneous QoS requirements of SUs can also be satisfied. In conclusion, by applying either RSSA or RMSA mechanisms, all users can be provided with enough economic incentives to participate in the spectrum sharing.

References

1. G. Wu, P. Ren, Q. Du, Recall-based dynamic spectrum auction with the protection of primary users. IEEE J. Sel. Areas Commun. 30(10), 2070–2081 (2012)
2. C. Yi, J. Cai, Multi-item spectrum auction for recall-based cognitive radio networks with multiple heterogeneous secondary users. IEEE Trans. Veh. Technol. 64(2), 781–792 (2015)
3. A.S. Alfa, *Queueing Theory for Telecommunications: Discrete Time Modelling of a Single Node System* (Springer, New York, 2010)
4. D. Xu, E. Jung, X. Liu, Efficient and fair bandwidth allocation in multichannel cognitive radio networks. IEEE Trans. Mob. Comput. 11(8), 1372–1385 (2012)
5. Y. Yuan, P. Bahl et al., KNOWS: cognitive radio networks over white spaces, in *Proceedings of IEEE DySPAN* (2007), pp. 416–427
6. L. Gao, Y. Xu, X. Wang, MAP: multiauctioneer progressive auction for dynamic spectrum access. IEEE Trans. Mob. Comput. 10(8), 1144–1161 (2011)
7. Z. Zheng, G. Chen, A strategy-proof combinatorial heterogeneous channel auction framework in noncooperative wireless networks. IEEE Trans. Mob. Comput. 14(6), 1123–1137 (2015)
8. C. Li, Z. Liu et al., Two dimension spectrum allocation for cognitive radio networks. IEEE Trans. Wirel. Commun. 13(3), 1410–1423 (2014)
9. R.B. Myerson, *Game Theory: Analysis of Conflict* (Harvard University Press, Cambridge, 1991)
10. A. Pekeč, M.H. Rothkopf, Combinatorial auction design. Manag. Sci. 49(11), 1485–1503 (2003)
11. A. Likhodedov, T. Sandholm, Methods for boosting revenue in combinatorial auctions, in *Proceedings of AAAI* (2004), pp. 232–237
12. D. Lehmann, L.I. Oćallaghan, Y. Shoham, Truth revelation in approximately efficient combinatorial auctions. J. ACM 49(5), 577–602 (2002)
13. V. Krishna, *Auction Theory* (Academic, New York, 2009)
14. D. Tse, P. Viswanath, *Fundamentals of Wireless Communication* (Cambridge University Press, Cambridge, 2005)
15. S. Martello, P. Toth, *Knapsack Problems: Algorithms and Computer Implementations* (Wiley, New York, 1990)
16. D. Niyato, E. Hossain, Market-equilibrium, competitive, and cooperative pricing for spectrum sharing in cognitive radio networks: analysis and comparison. IEEE Trans. Wirel. Commun. 7(11), 4273–4283 (2008)

Chapter 4
Two-Stage Spectrum Sharing Mechanism

4.1 Introduction

In this chapter, we consider a more complicated scenario of spectrum sharing with multiple spectrum sellers. In this model, a CR network with multiple heterogeneous POs and SUs is considered. Each PO has a different amount of spectrum to lease in different specific areas, and has a different users' (PUs') activity. Each SU has heterogeneous requirements in terms of spectrum demands and attitudes toward POs' potential spectrum recall. Obviously, in this case, spectrum sharing needs to jointly consider both spectrum allocation and individual strategies. However, solving such a joint optimization problem is challenging due to the facts that (1) PUs' activities are random and heterogenous among all POs; and (2) before the spectrum allocation has been done, it is impossible to know the quantity of spectrum recalled from each SU. In order to deal with the high computational complexity involved in solving such problem, we introduce a new method called *Two-stage resource allocation scheme with combinatorial auction and Stackelberg game in spectrum sharing (TAGS)* mechanism [1], which decomposes the solution into two separate stages. In the first stage, a suboptimal spectrum allocation is derived by formulating a combinatorial spectrum auction without considering the potential spectrum recall. Based on the winner determination in the first stage, each PO then decides a maximum amount of spectrum that may be recalled in the second stage, and each winning SU claims a payment reduction so as to offset the risk of utility degradation. Such a decision making process is viewed as a Stackelberg pricing game, and the best strategies for both POs and SUs are figured out accordingly. Theoretical and simulation results demonstrate that TAGS mechanism is efficient in increasing the spectrum utilization and economically feasible for all participants.

The rest of this chapter is organized as follows: Sect. 4.2 describes the system model and summarizes all important notations used in this chapter. Section 4.3 presents the first stage of TAGS mechanism, i.e., the combinatorial spectrum

© The Author(s) 2016
C. Yi, J. Cai, *Market-Driven Spectrum Sharing in Cognitive Radio*,
SpringerBriefs in Electrical and Computer Engineering,
DOI 10.1007/978-3-319-29691-3_4

auction. A recall-based pricing game is formulated in Sect. 4.4 to study the strategy decision process in the second stage of TAGS. Section 4.5 shows the analyses of some desired economic properties and a detailed time-line of the TAGS mechanism. Simulation results are illustrated in Sect. 4.6. Finally, a brief summary is given in Sect. 4.7.

4.2 System Model

Consider a CR network consisting of m POs and n SUs as shown in Fig. 4.1. Each PO i owns bandwidth $\mathcal{T}_w^i$ to serve its own subscribed PUs. Assume that PUs with the same PO i are homogeneous in terms of spectrum demand s^i and individual utility u^i (u^i could be set as the valuation of achievable rate by receiving s^i). However, PUs from different POs may be heterogeneous. If the remaining spectrum of PO i is less than s^i, newly arrived PUs have to wait in the queue and will be served later based on the FCFS rule. Without loss of generality, PUs arrive at each PO i following a Poisson process with an average arrival rate of λ_i. Furthermore, assume that the spectrum occupancy time of PUs in PO i is independent and identically exponentially distributed with service rate μ_i. Then, the service of PUs in PO i can be regarded as an M/M/c queueing system with $c = \lfloor \mathcal{T}_w^i/s^i \rfloor$.

Each PO has a predefined specific geographic region for spectrum marketing and lease a certain quantity of spectrum to SUs within this area, while at the same time guaranteeing the QoS for its own PUs. Consider the mean waiting time in the queue as the measurement of QoS for PUs. Then, if the mean waiting time of PUs in PO i, i.e., M_w^i, is longer than a certain requirement γ_i, PO i has to be punished for the QoS degradation.

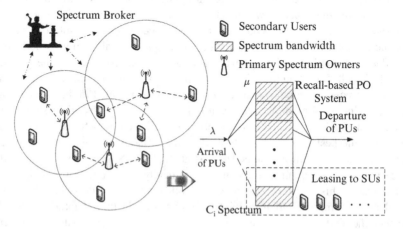

Fig. 4.1 System model of TAGS

Assume that all the POs are synchronized with same time frames, and spectrum sharing is carried out frame by frame. At the beginning of each frame, each PO determines the quantity of spectrum for leasing based on its own PUs' current spectrum usage. However, due to the random activities of PUs, POs may have insufficient spectrum to serve a sudden increasing demand from their own PUs if their unused spectrum has already been auctioned off. Since PUs are granted with higher spectrum access priority in CR networks, spectrum recall is enabled for POs, i.e., each PO can recall some auctioned spectrum from the winning SUs to satisfy its own PUs' demands if necessary. Note that POs are not necessary to recall spectrum for all newly arrived PUs. In fact, each PO can tolerate suffering from a degradation on PUs' QoS if its overall utility can be improved. Moreover, let us assume that recalled spectrum will not be returned to SUs until the next time frame. Certainly, winning SUs will get corresponding refunds if their winning spectrum were recalled by POs.

For simplicity, all SUs are assumed to be located within their interference ranges so that spectrum spatial reuse is not allowed. Furthermore, since it is difficult in employing discontinuous spectrum bands from different operators (POs) for a radio device with limited physical layer capability [2, 3], each SU can only access the spectrum from a same PO.

Consider that there is a small period used for spectrum management at the beginning of each frame. Since each PO predefines its specific region for spectrum leasing and SUs are randomly scattered in the entire area, it is difficult to find an optimal spectrum allocation in a distributed manner. Thus, a central entity, called *spectrum broker*, is introduced in the network. However, even with the central broker, it is still hard, if possible, to jointly determine the optimal spectrum allocation and best spectrum recall strategies because (1) the optimal amount of spectrum recall from each PO relies on a pre-existing optimal winner assignment; (2) the optimal winner assignment should be based on the optimal bids collected from SUs; and (3) the optimal bids are in turn determined regarding the potential spectrum recall. In order to tackle the complexity of this issue, a two-stage solution, i.e., TAGS, is introduced, which can be illustrated as in Fig. 4.2. In the first stage, each PO reports to the *spectrum broker* the quantity of spectrum for leasing and its specific spectrum trading region. Note that since the recall information is unknown in the first stage, the amounts of spectrum recalled by POs cannot be considered as strategies in the auction. At the same time, each SU sends out its private information, including its spectrum demand, bidding price, and location. The *broker* collects all these sealed bids and determines an optimal allocation which leads to a social optimality without considering the spectrum recall. In addition, the *broker* calculates the payments and payoffs for SUs and POs, respectively. In the second stage, each PO informs its own winning SUs a maximum quantity of spectrum which may be recalled, along with its spectrum recall scheme. Each SU then determines a reduction on its payment so that its utility can be maximized with such strategy. Finally, POs in turn derive the spectrum recall ratio on each winner.

For convenience, Table 4.1 lists some important notations used in this chapter.

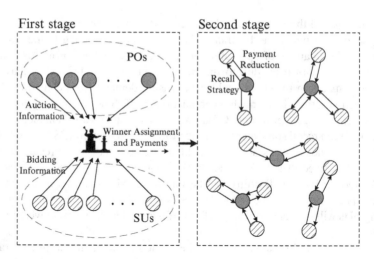

Fig. 4.2 An illustration of all actions in TAGS

Table 4.1 Important notations in this chapter

Notation	Meaning
m	Number of POs
n	Number of SUs
$\mathcal{T}_w^i$	Total spectrum bandwidth of PO i
γ_i	Mean waiting time requirement of PO i
C_i	Amount of auctioned spectrum offered by PO i
$\mathcal{E}_j$	Status indicator of SU j
p_j	Payment of SU j determined in the first stage
R_i	Payoff that each PO obtain after the auction
W_i	Set of winning SUs allocated to PO i
ω_i	Spectrum recall ratio determined by PO i
$\mathcal{RC}_i$	Maximum amount recalled by PO i
β_j	Payment reduction parameter of winning SU j
r_j	Quantity of spectrum recall on SU j

4.3 First Stage: Combinatorial Spectrum Auction

In this section, a centralized combinatorial spectrum auction is introduced that will be adopted in the first stage of TAGS. The corresponding winner determination problem (WDP) is first formulated by binary integer programming (BIP). Due to the computational intractability of BIP, a novel combinatorial auction mechanism is then applied, which can run in a polynomial time.

4.3.1   Winner Determination Problem

The defined combinatorial spectrum auction consists of m sellers (POs), each with heterogeneous amount of goods (spectrum) to sell, and n buyers (SUs) with different demands. Define the set of POs as $\mathcal{M}$ with $|\mathcal{M}| = m$. Each PO i determines the quantity of spectrum for leasing, called $C_i \in \mathbb{R}$, based on its own PUs' activities. For simplicity, the spectrum trading area of each PO i is assumed to be a circle in an Euclidean plane with the location of this PO as the center and the radius of R_i.

Similarly, let us define $\mathcal{N}$ as the set of SUs with $|\mathcal{N}| = n$. Each SU j has a specific spectrum demand, $D_j \in \mathbb{R}$, and a private valuation for its demand, v_j. Without loss of generality, let v_j equal the monetary value of Shannon capacity that SU j can obtain over D_j spectrum bandwidth as

$$v_j = \sigma_j D_j \log_2(1 + \eta_j), \tag{4.1}$$

where σ_j and η_j are the monetary weight index and the SNR, respectively. Both σ_j and η_j are supposed to be constant for each SU j [4].

At the beginning of the auction, POs report their auction information $A_i, i \in \mathcal{M}$, to the *spectrum broker*. Here, each A_i is a 3-tuple (C_i, O_i, L_i), where

- C_i is the spectrum bandwidth provided by PO i for leasing.
- O_i represents the location of seller i, which is assumed to be a coordinate in Euclidean plane, i.e., $O_i = (o_i^x, o_i^y)$.
- L_i denotes the radius of PO i's spectrum trading region.

Meanwhile, SUs send their bids to the *spectrum broker*, denoted as $B_j, j \in \mathcal{N}$. Each bid B_j is also specified as a 3-tuple (D_j, b_j, l_j), where

- D_j is the spectrum demand of buyer j.
- b_j indicates the amount that the buyer is willing to pay for D_j. Note that for truthful auctions, the bidding price equals the true valuation, i.e., $b_j = v_j$.
- l_j represents the location of buyer j, which is assumed to be a coordinate in Euclidean plane, i.e., $l_j = (l_j^x, l_j^y)$.

After receiving all these information, the *spectrum broker* first identifies the locations of all buyers and sellers, and then groups the buyers into m sets according to the spectrum trading area of each seller. Note that these m sets may be overlapped, i.e., each SU could be located in multiple POs' trading areas. Let Y_i denote the set of buyers who locate in the auction coverage of PO i. Obviously, all buyers in set Y_i should satisfy the following condition:

$$\sqrt{(l_j^x - o_i^x)^2 + (l_j^y - o_i^y)^2} \leq L_i, \quad \forall j \in Y_i. \tag{4.2}$$

After grouping the buyers, the *spectrum broker* formulates an optimization problem to determine the winners in order to maximize the social welfare, i.e., the total bidding price from all winning SUs. The formulated optimization problem is

$$\max_{\{x_{ij}, \forall i \in \mathcal{M}, \forall j \in \mathcal{N}\}} \sum_{i=1}^{m} \sum_{j=1}^{n} b_j x_{ij} \tag{4.3}$$

$$s.t. \quad \sum_{j \in Y_i} D_j x_{ij} \leq C_i, \ \forall i \in \mathcal{M}, \tag{4.4}$$

$$\sum_{i=1}^{m} x_{ij} \leq 1, \qquad \forall j \in \mathcal{N}, \tag{4.5}$$

$$x_{ij} \in \{0, 1\}, \qquad \forall i \in \mathcal{M}, \ \forall j \in \mathcal{N}. \tag{4.6}$$

where $x_{ij} = 1$ if SU j is allocated to PO i, and $x_{ij} = 0$, otherwise.

The first constraint means that for winning SUs allocated to PO i, their total demands should be less than or equal to the quantity of spectrum that PO i offers. The second constraint limits each SU to access spectrum from no more than one PO. The third constraint assumes that all buyers are single-minded so that there are only two outcomes for each SU, i.e., win or lose. If the optimal winner assignment can be obtained, then the well-known VCG mechanism can be applied to calculate the payments in order to ensure both economic robustness and efficiency of the auction. However, the above WDP is obviously a BIP which can be proved as NP-hard by reducing it to a weighted independent set problem [5]. Therefore, it is necessary to explore an alternative mechanism with an approximate winner determination algorithm and a tailored payment scheme.

4.3.2   Polynomial-Time Approximation Algorithm for the WDP

The following algorithm is inspired by the approximation algorithm for multiple knapsack problem (MKP) in [6]. However, different from the general MKP, where items could be allocated to any knapsack without considering geographic restriction, the WDP algorithm has to take both the POs' auction region and SUs' locations into account.

After receiving all the bids, according to the idea of *LOS ordering* shown in Chap. 2, the *spectrum broker* first sorts the SUs based on a decreasing order of $b_j / \sqrt{D_j} \ j = 1, \ldots, n$, and sorts POs based on an increasing order of the amount of auctioned spectrum, $C_i, i = 1, \ldots, m$, i.e.,

$$\frac{b_1}{\sqrt{D_1}} \geq \frac{b_2}{\sqrt{D_2}} \geq \ldots \geq \frac{b_j}{\sqrt{D_j}} \geq \ldots \geq \frac{b_n}{\sqrt{D_n}}; \tag{4.7}$$

$$C_1 \leq C_2 \leq \ldots \leq C_i \leq \ldots \leq C_m. \tag{4.8}$$

Notice that, although it is more straightforward to order the SUs with their unit bidding price, i.e., $b_j / D_j, \forall j \in \mathcal{N}$, such order may undervalue bids with large

---

Algorithm 2 Initial solution for WDP

1: **input:** n, b_j, D_j, l_j m, C_i, O_i, L_i;
2: **output:** $z, \mathcal{E}_j$;
3: $z = 0$;
4: **for** $j = 1$ to n **do**
5:    $\mathcal{E}_j = 0$;
6: **end for**
7: **for** $i = 1$ to m **do**
8:    $\overline{C}_i = C_i, Y_i = \emptyset$;
9:    **for** $j = 1$ to n **do**
10:       **if** $\sqrt{(l_j^x - o_i^x)^2 + (l_j^y - o_i^y)^2} \leq L_i$ **then**
11:         $Y_i = Y_i \cup \{j\}$;
12:       **end if**
13:       **for** each $j \in Y_i$ **do**
14:         **if** $\mathcal{E}_j = 0$ **and** $D_j \leq \overline{C}_i$ **then**
15:           $\mathcal{E}_j = i, \overline{C}_i = \overline{C}_i - D_j, z = z + b_j$;
16:         **end if**
17:       **end for**
18:    **end for**
19: **end for**

---

demands. Recall that in Chap. 2, it has been proved that ordering buyers with $b_j/\sqrt{D_j}$ can provide a better approximation ratio to optimality. In addition, in (4.7) and (4.8), all the indices are rearranged and the following searching procedure will follow this order.

The initial feasible solution can be first generated by using Algorithm 2. Define z as the overall bidding price of the auction, $\overline{C}_i$ as the remaining capacity (in terms of the spectrum bandwidth) of knapsack (PO) i and $\mathcal{E}_j$ as the status indicator of SU j, where

$$
\mathcal{E}_j = \begin{cases} 0, & \text{if SU } j \text{ is currently unallocated;} \\ \text{index of the PO it is allocated to,} & \text{otherwise.} \end{cases}
$$

Algorithm 2 considers the POs one by one. For any PO i, the algorithm assigns the spectrum to the unallocated SUs which are covered in the auction region of PO i until the remaining capacity is smaller than the request from any unallocated SUs. For each feasible allocation, the algorithm updates the following parameters as $\mathcal{E}_j = i$ and $z = z + b_j$.

Subsequently, the initial solution is improved by local exchanges [7]. The improvement consists of three processes, i.e., rearrangement, interchange, and replacement:

- *Rearrangement*
  Consider all SUs with $\mathcal{E}_j > 0$ according to the increasing order of $b_j/\sqrt{D_j}$. Rearrange these SUs one by one to its next potential trading PO with sufficient

remaining capacity in a cyclic manner, i.e., in the order of $\{\mathcal{E}_j + 1, \mathcal{E}_j + 2, \ldots, m, 1, 2, \ldots, \mathcal{E}_j - 1\}$. Note that, with this rearrangement, the SUs with less demand may be assigned to the PO with smaller residual capacities so that more capacity in the current PO may be available to unallocated SUs.

- *Interchange*

  The interchange process considers all pairs of allocated SUs and, if possible, interchanges their PO assignment whenever doing so allows insertion of a new SU to one of the knapsacks (POs). Through this algorithm, social welfare (the value of z) can be enhanced since the number of winning SUs is increased.

- *Replacement*

  This process aims to replace any already allocated SU by one or more unallocated SUs which are also covered in the trading area of the same PO, so that the total profit is increased.

Based on the analysis in [7], it is not difficult to prove that no step of these algorithms needs more than $O(n^2)$ time. Thus, an approximately optimal solution of the WDP can finally be obtained by sequentially executing the algorithms presented above in polynomial time.

4.3.3 Payment Design

Since the VCG payment rule is incompatible with approximate WDP algorithm in a quite general sense, the idea of *LOS pricing scheme* is adopted here for determining the charging prices that are "Vickrey-like." Specifically, the payment of each winning SU j should be a function of the highest-value bid that j's bid *blocks*.

Definition 4.1 (Blocks). Suppose bid B_j was granted by the WDP algorithm while bids in set $\mathbf{B_j}-$ were denied. The bid B_j *blocks* $\mathbf{B_j}-$ if, after removing the bid B_j from the auction, all bids in $\mathbf{B_j}-$ would be granted.

Based on this definition, the payment of each SU j can be calculated by distinguishing two cases:

- If SU j loses or it wins but *blocks* no other bid (i.e., $\mathbf{B_j}- = \emptyset$), then its payment is 0.
- If SU j is granted its demand D_j and $\mathbf{B_j}- \neq \emptyset$, the payment p_j of SU j is set as

$$p_j = \sqrt{D_j} \times \max_{k \in \mathbf{B_j}-} \left(\frac{b_k}{\sqrt{D_k}} \right). \tag{4.9}$$

After deciding the charges from all the winning SUs, the *spectrum broker* is responsible to determine the payoffs to each PO based on the number of SUs allocated to it. The income of PO i, R_i, can be easily derived as

$$R_i = \sum_{j \in Y_i} x_{ij} p_j = \sum_{j=1}^{n} x_{ij} p_j. \tag{4.10}$$

Note that both $p_j, j \in \mathcal{N}$ and $R_i, i \in \mathcal{M}$ are considered as the contract made by the first-stage spectrum allocation. All the users would follow this contract along with its corresponding spectrum allocation and bring them into the next stage.

4.3.4 Auction Properties

Here, we present the proofs for individual rationality and incentive compatibility of the first-stage auction.

Lemma 4.1. *The auction mechanism in the first stage provides individual rationality for all truthful buyers (i.e., $b_j = v_j$).*

Proof. The utility of SU j is zero if it loses the auction. Otherwise, the utility of winning SU j can be calculated as

$$U_j^s = v_j - p_j = b_j - \sqrt{D_j} \times \max_{k \in B_j^-} \left(\frac{b_k}{\sqrt{D_k}} \right) = \left[\frac{b_j}{\sqrt{D_j}} - \max_{k \in B_j^-} \left(\frac{b_k}{\sqrt{D_k}} \right) \right] \times \sqrt{D_j} \geq 0.$$

The above inequality holds since SU j is a winner, and thus $\frac{b_j}{\sqrt{D_j}} \geq \max_{k \in B_j^-} \left(\frac{b_k}{\sqrt{D_k}} \right)$ according to Definition 4.1. Hence, the payment scheme in the first stage can ensure non-negative utilities. □

Lemma 4.2. *The auction mechanism in the first stage is incentive-compatible which means that no buyer could obtain higher utility by bidding untruthfully.*

Proof. Two different cases should be considered to prove this lemma:

Case I: SU j wins the auction and gets utility $U_j^s \geq 0$ when bidding truthfully. If SU j bids untruthfully ($b_j' \neq v_j$), there are two possible outcomes, i.e., (i) SU j loses the auction and gets $U_j^s = 0$; or (ii) SU j still wins the auction and its utility becomes

$$\widehat{U_j^s} = v_j - p_j' = v_j - \sqrt{D_j} \times \max_{k \in B_j^-} \left(\frac{b_k}{\sqrt{D_k}} \right) \tag{4.11}$$

Obviously, we have $\widehat{U_j^s} = U_j^s$.

Case II: SU j loses the auction when bidding truthfully and get utility $U_j^s = 0$. Its utility may be changed only if SU j wins with an untruthful bidding. Let b_j and b_j' denote truthful bidding and untruthful bidding, respectively. We have

$\frac{b'_j}{\sqrt{D_j}} \geq \max_{k\in B'_{j-}} (\frac{b_k}{\sqrt{D_k}}) \geq \frac{b_j}{\sqrt{D_j}}$, otherwise SU j still cannot win the auction. In this case, its utility can be proved to be non-positive:

$$\widehat{U^s_j} = v_j - p'_j = v_j - \sqrt{D_j} \times \max_{k\in B'_{j-}} \left(\frac{b_k}{\sqrt{D_k}} \right)$$

$$\leq v_j - \sqrt{D_j} \times \frac{b_j}{\sqrt{D_j}} = v_j - b_j = 0$$

(4.12)

In summary, SU j cannot increase its utility by bidding any other value than v_j. In other words, bidding truthfully is a dominant strategy for each buyer.   □

4.4 Second Stage: Recall-Based Pricing Game

Given the auction output from the first stage, the impacts of spectrum recall on both POs' and SUs' utilities are investigated in the second stage. The strategy decision process is formulated as a Stackelberg game, and the Nash equilibrium (NE) of such game is analyzed accordingly.

4.4.1 Stackelberg Game Formulation

All POs and SUs are assumed to be intelligent in the considered framework. Each PO first announces a maximum quantity of spectrum that may be recalled, and then each SU determines a payment reduction in order to maximize its utility under this recall-based system.

The strategy decision process can be formulated as a Stackelberg game, in which POs act as leaders and SUs play as followers. The leader selects the optimal strategy based on the knowledge of its effect on the followers' actions. For winning SUs, they have been assigned their desired spectrum in the first stage. However, they are informed of a potential spectrum recall by their allocated POs. Thus, SUs within one PO would compete with each other in order to decrease the amount of recalled spectrum from themselves. This results in a non-cooperative payment reduction game, where pricing scheme can be used to adjust the amount of spectrum recalled on SUs according to their payments.

The NE of this recall-based pricing game can be solved by backward induction. Namely, the NE of the game among SUs can be first derived given the quantity of recalled spectrum, and then the best responses of POs can be calculated.

4.4.2   Utilities of Primary Spectrum Owners and Secondary Users

Each PO and its assigned winning SUs are formed in one group after the winner assignment in the first stage. Thus, each PO can run its Stackelberg game independently in its own group, and its spectrum recall strategy would not affect the other users in other groups. From this observation, we can focus on one PO only and omit the subscript i in the following context for notation simplicity.

4.4.2.1   Recall Scheme

Let W be the set of winning SUs allocated to a PO. Then, the total spectrum bandwidth assigned to W after the first stage is $T_w^T = \sum_{j \in W} D_j$. Let us define that the maximum quantity of spectrum recalled from W declared by the PO is

$$\mathcal{RC} = \omega T_w^T = \omega \sum_{j \in W} D_j, \qquad \omega \in [0, 1], \tag{4.13}$$

where ω represents the percentage of spectrum recalled from T_w^T, and is declared by the PO at the beginning of the second stage.

Thus, the PO actually reserves $W - (1 - \omega)T_w^T$ spectrum for its own users. Though the PO will compensate SUs in W for the spectrum recall, this behavior violates the contract made in the first stage. As a response, each winning SU j can determine a parameter $\beta_j \in [0, 1]$, so that the actual payment of SU j is reduced to $\beta_j p_j$. Let r_j be the quantity of spectrum recalled from SU j. Then, if $r_j = D_j$, all winning spectrum of SU j would be recalled so that its utility turns to be zero or $\beta_j = 0$. By considering the fairness on spectrum recall, let us define r_j as

$$r_j = \mathcal{RC} \times \left(\frac{1/\varepsilon_j}{\sum_{k \in W} 1/\varepsilon_k} \right), \qquad \forall j \in W, \tag{4.14}$$

where ε_j indicates the actual unit payment of SU j and can be calculated as

$$\varepsilon_j = \frac{\beta_j p_j}{D_j}, \qquad \forall j \in W. \tag{4.15}$$

According to the definitions of (4.14) and (4.15), r_j is inversely proportional to the actual unit payment declared by each SU. Assume that $D_j \geq r_j, \forall j \in W$, so that all winning SUs are willing to follow this spectrum recall scheme. In fact, this assumption can be relaxed unless the total quantity of spectrum recall is larger than a certain threshold Γ (which will be further explained in Sect. 4.5). Furthermore, the recall scheme defined in (4.14) is regarded as a common knowledge to all users.

4.4.2.2 Utility Function of Each Winning SU

With spectrum recall, the transmission rate of SU j can be rewritten as

$$G_j = (D_j - r_j) \log_2(1 + \eta_j), \quad \forall j \in W, \tag{4.16}$$

where $D_j - r_j$ denotes the bandwidth that SU j can actually obtain through its final payment $\beta_j p_j$.

In this recall-based system, the goal of all winning SUs is to prevent their transmission rates from experiencing significant degradation, while at the same time, lower their payments in order to reduce the risk from spectrum recall. Obviously, (4.16) is only a function of r_j because all other parameters are fixed. Since r_j is decided by the parameter β_j, the strategy of SU j is actually the determination of β_j. The change on β_j would ultimately lead to an impact on PO's spectrum recall distribution on its winners, and in turn determine the achievable rate of SU j.

POs may recall their leased spectrum to deal with their own users' demand peak during the transmission period. Thus, the services of winning SUs would be degraded. The compensation for such degradation can be formulated as

$$H_j = \varsigma_j r_j, \tag{4.17}$$

where ς_j represents the reported compensation index which can be calculated by SU j as

$$\varsigma_j = \kappa_j \log_2(1 + \eta_j). \tag{4.18}$$

Here κ_j is defined as a coefficient of compensation rate for SU j and it also reflects the SU's attitude toward spectrum recall. Note that κ_j is a parameter pre-determined by the system based on the user's service requirement, so that it cannot be changed arbitrarily in the game.

With all above settings, the utility function of SU j, i.e., U_j^s, can be formulated, which includes the benefit through its achievable transmission rate, actual payment, and the monetary compensation for spectrum recall. The expression of U_j^s can be expressed as

$$U_j^s = \sigma_j G_j(\beta_j) - \beta_j p_j + H_j, \tag{4.19}$$

where $\sigma_j G_j(\beta_j)$ is the valuation of the transmission rate actually obtained by SU j, and it is also a function of the variable β_j. Substitute (4.17) and (4.18) into (4.19), the utility function of SU j can be rewritten as

$$U_j^s = [\sigma_j D_j - (\sigma_j - \kappa_j) r_j] \log_2(1 + \eta_j) - \beta_j p_j, \tag{4.20}$$

where $\sigma_j - \kappa_j \geq 0$, which indicates that U_j^s should not be increased with the quantity of recalled spectrum, so that each winning SU would compete with others for reducing its r_j.

4.4.2.3 Utility Function of the PO

For the PO, let us denote the spectrum bandwidth available for its own PUs as $\mathcal{S}$. Obviously, $\mathcal{S}$ consists of the quantity of both unleased and recalled spectrum, i.e., $\mathcal{S} = \mathcal{T}_w - \mathcal{T}_w^T + \mathcal{R}\mathcal{C}$. Thus, the maximum number of PUs that PO can accommodate is

$$\mathcal{F} = \left\lfloor \frac{\mathcal{S}}{s} \right\rfloor = \left\lfloor \frac{\mathcal{T}_w - (1 - \omega) \sum_{j \in W} D_j}{s} \right\rfloor, \tag{4.21}$$

where s is the spectrum demand of each PU. By considering $\mathcal{F}$ as the number of servers in an M/M/c queueing system, i.e., $c = \mathcal{F}$, the mean waiting time for arriving PUs can be calculated as

$$M_w = \frac{\mathcal{Q}(\mathcal{F}, \zeta)}{\mu(\mathcal{F} - \zeta)}, \tag{4.22}$$

where $\zeta = \lambda/\mu$ denotes the utilization factor and $C(\mathcal{F}, \zeta)$ is the queueing probability as

$$\mathcal{Q}(\mathcal{F}, \zeta) = \frac{\zeta^{\mathcal{F}}/\mathcal{F}!}{[(\mathcal{F} - \zeta)/\mathcal{F}] \sum_{k=0}^{\mathcal{F}-1}(\zeta^k/k!) + \zeta^{\mathcal{F}}/\mathcal{F}!}. \tag{4.23}$$

According to the considered model, PUs has a QoS requirement that the mean waiting time should be not longer than γ, otherwise a penalty will be introduced in the utility function of the PO. However, the ultimate goal of PO is to maximize its total utility including the benefit from its own users' service and the economic revenue from spectrum leasing. Therefore, depending on the penalty and profit, the PO may not always try to keep M_w being less than γ so as to maximize its overall utility.

If $\omega = 0$, the PO will not enter the second stage since there is no need to build a recall-based pricing game. Hence, only the case when $\omega \in (0, 1]$ and $M_w \geq \gamma$ has to be considered. In this scenario, the PO's benefit gained from its own users, χ, can be expressed as

$$\chi = \mathcal{F}u - \Lambda(\psi, M_w, \gamma), \tag{4.24}$$

where u is the individual utility of the PU, $\Lambda(\cdot)$ denotes the penalty function, and ψ represents the weight index of the penalty for QoS degradation.

The utility of PO consists of three terms, i.e., benefit from its PUs, profits from spectrum leasing, and the compensation caused by its spectrum recall. Mathematically, the utility function of PO can be presented as

$$U^p = \chi(\omega) + \sum_{j \in W} \beta_j p_j - \sum_{j \in W} H_j, \qquad (4.25)$$

Therefore, each PO will try to maximize its own utility by choosing the best strategy of ω.

4.4.3 Nash Equilibrium of the Game Among Secondary Users

Given ω, SUs in W will compete with each other to maximize their utilities by selecting their own strategy β. Let t denote the number of SUs in set W, i.e., $|W| = t$, and rearrange the indices of these SUs from 1 to t. Then, a non-cooperative pricing game can be denoted as $\mathcal{G}_{su} = \{t, \{\hat{\beta}_j\}, \{U_j^s(\cdot)\}\}$, where $\hat{\beta}_j$ and $U_j^s(\cdot)$ are the strategy set and the utility function of SU j, respectively. We further assume that $t > 1$ (otherwise there is no competition among SUs or the game does not exist).

In this game, each SU j, $1 \leq j \leq t$, selects its strategy β_j to maximize its utility $U_j^s(\beta_j, \boldsymbol{\beta}_{-j})$, where $\boldsymbol{\beta}_{-j}$ represents the strategies of all other SUs in W. Then, the NE of the game can be defined as follows.

Definition 4.2. A strategy profile $\boldsymbol{\beta} = (\beta_1, \beta_2, \ldots, \beta_t)$ is an NE of the game $\mathcal{G}_{su} = \{t, \{\hat{\beta}_j\}, \{U_j^s(\cdot)\}\}$ if for every user j, $U_j^s(\beta_j, \boldsymbol{\beta}_{-j}) \geq U_j^s(\beta_j', \boldsymbol{\beta}_{-j})$ for all $\beta_j' \in \hat{\beta}_j$.

Next, the existence and uniqueness of NE are proved, and then the unique NE point of the game is calculated.

4.4.3.1 Existence of NE

Theorem 4.1. The game $\mathcal{G}_{su} = \{t, \{\hat{\beta}_j\}, \{U_j^s(\cdot)\}\}$ has at least one NE.

Proof. Since $\beta_j \in [0, 1]$, it is obvious that $\hat{\beta}_j$ is a nonempty, convex, and compact subset of the Euclidean space $\mathbb{R}^n$.

According to the utility function of SU j in (4.20), it is not difficult to find that U_j^s is continuous. Taking the first order derivative of U_j^s with respect to β_j, we have

$$\frac{\partial U_j^s}{\partial \beta_j} = -(\sigma_j - \kappa_j) \log_2(1 + \eta_j) \frac{\partial r_j}{\partial \beta_j} - p_j. \qquad (4.26)$$

With the definition of r_j in (4.14) and (4.15), we have

$$\frac{\partial r_j}{\partial \beta_j} = -\frac{\mathcal{RC} \cdot J_{1,j} J_{2,j}}{(1 + J_{1,j} J_{2,j} \beta_j)^2}, \qquad (4.27)$$

where $J_{1,j} = p_j/D_j$ and $J_{2,j} = \sum_{k \neq j} \frac{D_k}{\beta_k p_k}$.

By substituting (4.27) to (4.26), the second order derivative of U_j^s can be calculated as

$$\frac{\partial^2 U_j^s}{\partial \beta_j^2} = -(\sigma_j - \kappa_j) \log_2(1 + \eta_j) \frac{\mathcal{RC}(J_{1,j}J_{2,j})^2}{(1 + J_{1,j}J_{2,j}\beta_j)^3} \le 0. \tag{4.28}$$

The above inequality holds since $\sigma_j \ge \kappa_j$ and no other terms are less than 0. Hence, the second derivative of U_j^s is always less than or equal to 0, which means that $U_j^s(\cdot)$ is concave in its strategy space. Thus, according to [8], $\mathcal{G}_{su}$ has at least one NE since the game $\mathcal{G}_{su}$ has a nonempty, convex and compact strategy space $\hat{\beta}_j$, and $U_j^s(\cdot)$ is continuous and concave.   □

4.4.3.2   Uniqueness of NE

Theorem 4.2. *The game $\mathcal{G}_{su} = \{t, \{\hat{\beta}_j\}, \{U_j^s(\cdot)\}\}$ has an unique NE.*

Proof. Let $\delta_j(\beta)$ be the best response function of SU j. We can first check whether $\delta_j(\beta)$ is a standard function.

Given the utility function of SU j in (4.20), the best response $\delta_j(\beta)$ can be obtained by solving the following equation:

$$\frac{\partial U_j^s}{\partial \beta_j} = (\sigma_j - \kappa_j) \log_2(1 + \eta_j) \frac{\mathcal{RC} \cdot J_{1,j}J_{2,j}}{(1 + J_{1,j}J_{2,j}\beta_j)^2} - p_j = 0. \tag{4.29}$$

To simplify the formula, let $J_{3,j} = (\sigma_j - \kappa_j)\mathcal{RC} \log_2(1 + \eta_j)$. Then, the solution of (4.29) can be derived as

$$\delta_j(\beta) = \sqrt{\frac{J_{3,j}}{p_j J_{1,j}J_{2,j}}} - \frac{1}{J_{1,j}J_{2,j}}. \tag{4.30}$$

Obviously, the term $\frac{J_{3,j}}{p_j J_{1,j}J_{2,j}}$ is greater than 0. Furthermore, since the strategy space is defined in $[0, 1]$, the existence of solution requires the satisfaction of the following constraint:

$$1 \le \frac{J_{1,j}J_{2,j}J_{3,j}}{p_j} \le (J_{1,j}J_{2,j} + 1)^2. \tag{4.31}$$

Since $\delta_j(\beta)$ is obviously a quadratic function, $\delta_j(\beta)$ will be monotonically increasing when $\frac{\partial \delta_j(\beta)}{\partial \beta} \ge 0$. Let us first express this derivative as

$$\frac{\partial \delta_j(\beta)}{\partial \beta} = \frac{\partial J_{2,j}}{\partial \beta} \left(-\frac{1}{2} \sqrt{\frac{J_{3,j}}{p_j J_{1,j}J_{2,j}^2}} + \frac{1}{J_{1,j}J_{2,j}^2} \right). \tag{4.32}$$

From (4.31), we have

$$J_{2,j} \geq \sqrt{\frac{4p_j}{J_{1,j}J_{3,j}}}, \qquad (4.33)$$

Then, it will be easy proved that $\frac{\partial \delta_j(\beta)}{\partial \beta} \geq 0$, i.e., $\delta_j(\beta)$ is indeed a monotonically increasing function.

Furthermore, we can evaluate $\Phi \delta_j(\beta) - \delta_j(\Phi \beta)$ as

$$\Phi \delta_j(\beta) - \delta_j(\Phi \beta) = \Phi \times \frac{\sqrt{\frac{J_{1,j}J_{2,j}J_{3,j}}{p_j}} - 1}{J_{1,j}J_{2,j}} - \frac{\sqrt{\frac{J_{1,j}J_{2,j}J_{3,j}}{\Phi p_j}} - 1}{J_{1,j}J_{2,j}\frac{1}{\Phi}}$$

$$= (\Phi - \sqrt{\Phi}) \sqrt{\frac{J_{3,j}}{J_{1,j}J_{2,j}p_j}}. \qquad (4.34)$$

For $\forall \Phi > 1$, we always have $\Phi - \sqrt{\Phi} > 0$. Hence, the above equation is always positive, which means that $\Phi \delta_j(\beta) - \delta_j(\Phi \beta) > 0$ and $\delta(\beta)$ is scalable.

Since the best-response $\delta_j(\beta)$ is proved to be positive, monotonic, and scalable, according to [9], it is a standard function. From [10], we know that the game $\mathcal{G}_{su}$ with $\delta_j(\beta)$ as a standard function has a unique NE. □

4.4.3.3　The NE Point of the Game $\mathcal{G}_{su}$

For a non-cooperative game, NE is defined as the operation point(s) at which no player could improve the utility by changing its strategy unilaterally. Since the NE of the game $\mathcal{G}_{su}$ has been proved to be existing and unique, the unique NE point β_j^* can be derived by directly solving the following equation set [8]:

$$\beta_j^* = \sqrt{\frac{J_{3,j}}{p_j J_{1,j} \sum_{k \neq j} \frac{D_k}{p_k \beta_k^*}} - \frac{1}{J_{1,j} \sum_{k \neq j} \frac{D_k}{p_k \beta_k^*}}}, \quad \forall j \in W. \qquad (4.35)$$

Although the above equations are not difficult to be solved, deriving a closed-form expression is not easy. Since the best response of the leader (PO), i.e., ω^*, can only be obtained by substituting the NE point of the $\mathcal{G}_{su}$ into the PO's utility function, we need to express β_j^* in terms of ω as follows.

Consider a special case with only two winners, i.e., $|W| = t = 2$. According to (4.35), the equation set for the NE point becomes

$$\begin{cases} \beta_1^* = \sqrt{\frac{\mathcal{B}_1}{\mathcal{A}_1 \frac{1}{\mathcal{A}_2 \beta_2^*}} - \frac{1}{\mathcal{A}_1 \frac{1}{\mathcal{A}_2 \beta_2^*}}}, \\ \beta_2^* = \sqrt{\frac{\mathcal{B}_2}{\mathcal{A}_2 \frac{1}{\mathcal{A}_1 \beta_1^*}} - \frac{1}{\mathcal{A}_2 \frac{1}{\mathcal{A}_1 \beta_1^*}}}, \end{cases}$$

where $\mathcal{A}_j = J_{1,j}$ and $\mathcal{B}_j = J_{3,j}/p_j$.

After some simple manipulations, we have

$$
\begin{cases}
\frac{\beta_1^*}{A_2} + \frac{\beta_2^*}{A_1} = \sqrt{\frac{B_1 \beta_2^*}{A_1 A_2}}, \\
\frac{\beta_2^*}{A_1} + \frac{\beta_1^*}{A_2} = \sqrt{\frac{B_2 \beta_1^*}{A_2 A_1}}.
\end{cases}
$$

By simple observation, we could find out that β_1^* and β_2^* satisfy

$$
\beta_2^*/\beta_1^* = B_1/B_2. \tag{4.36}
$$

Therefore, the NE is

$$
\begin{cases}
\beta_1^* = \frac{A_1 A_2 B_1^2 B_2}{(A_1 B_1 + A_2 B_2)^2}, \\
\beta_2^* = \frac{A_1 A_2 B_1 B_2^2}{(A_1 B_1 + A_2 B_2)^2}.
\end{cases}
$$

Since $J_{3,j} = (\sigma_j - \kappa_j)\omega T_w^T \log_2(1 + \eta_j)$, we have $B_j = C_j \omega$, where $C_j = (1/p_j)$ $(\sigma_j - \kappa_j)T_w^T \log_2(1 + \eta_j)$. Thus,

$$
\begin{cases}
\beta_1^* = \frac{A_1 A_2 C_1^2 C_2}{(A_1 C_1 + A_2 C_2)^2} \cdot \omega, \\
\beta_2^* = \frac{A_1 A_2 C_1 C_2^2}{(A_1 C_1 + A_2 C_2)^2} \cdot \omega.
\end{cases}
$$

For $t > 2$, because of the symmetry property of the equation set, the general solution can be represented as

$$
\beta_j^* = \Pi_j \omega, \qquad \forall j \in W, \tag{4.37}
$$

where Π_j is a coefficient associated with SU j. For example, when $t = 2$, $\Pi_1 = \frac{A_1 A_2 C_1^2 C_2}{(A_1 C_1 + A_2 C_2)^2}$ and $\Pi_2 = \frac{A_1 A_2 C_1 C_2^2}{(A_1 C_1 + A_2 C_2)^2}$.

4.4.4 Best Response of the Primary Spectrum Owner

Based on the strategies made by the followers (SUs), the leader (PO) can then calculate its best response ω^* as follows.

By substituting (4.37) into (4.25), U^p can be expressed as

$$
U^p = \chi(\omega) - \omega \sum_{j \in W} \kappa_j T_w^T \log_2(1 + \eta_j) \frac{\frac{D_j}{\Pi_j p_j}}{\sum_{j \in W} \frac{D_j}{\Pi_j p_j}} + \omega \sum_{j \in W} \Pi_j p_j. \tag{4.38}
$$

Let $\vartheta = \sum_{j \in W} \kappa_j T_w^T \log_2(1 + \eta_j) \frac{\frac{D_j}{\Pi_j p_j}}{\sum_{j \in W} \frac{D_j}{\Pi_j p_j}} - \sum_{j \in W} \Pi_j p_j$. Then, (4.38) can be rewritten as

$$U^p = \chi(\omega) - \omega\vartheta. \tag{4.39}$$

Unfortunately, directly calculating the derivative of $\chi(\omega)$ with respect to ω is difficult, because it is hard to build the penalty function, $\Lambda(\cdot)$, based on (4.22) and (4.23). For explanation purpose, $\Lambda(\cdot)$ is simplified as a sigmoid function of ω. In fact, as ω increases, each PO can recall more spectrum to serve its own PUs so that the waiting time of arriving PUs will decrease. Thus, $\Lambda(\cdot)$ should be a decreasing function of ω. However, such decreasing trend should not be linear. Intuitively, when the amount of spectrum reserved by the PO is much less than the required amount of spectrum to ensure the desired QoS, the increase of ω will result in a significant improvement on the QoS. However, when the amount of reserved spectrum is close to the required amount, the effect of increasing ω becomes gradually. Obviously, such observation can be well depicted by a sigmoid function. Note that the sigmoid function has been widely used in the literature to formulate users' satisfaction with respect to service quality or resource allocation [11–13]. Specifically, the definition of $\Lambda(\cdot)$, i.e., PUs' degree of dissatisfaction to their QoS, can be defined as

$$\Lambda(\psi, M_c, \gamma, \omega) = \frac{1}{1 + e^{\psi^{-1}\omega(M_c - \gamma)}}, \tag{4.40}$$

where γ denotes the QoS requirement of its PUs, ψ indicates the weight index of penalty for QoS degradation, and M_c is the current waiting time of the system when spectrum recall is not enabled. Obviously, $\Lambda(\psi, M_c, \gamma, \omega)$ is a decreasing function of ω. Now, the function $\chi(\omega)$ can be rewritten as

$$\chi(\omega) = \mathcal{F}(\omega)u - \frac{1}{1 + e^{\psi^{-1}\omega(M_c - \gamma)}}, \tag{4.41}$$

where $\mathcal{F}(\omega)$ is the number of PUs that the PO can accommodate under such circumstance.

Let $v = \psi^{-1}(M_c - \gamma)$ and take the first order derivative of U^p in (4.39) with respect to ω. We have

$$\frac{\partial U^p}{\partial \omega} = \frac{v e^{v\omega}}{(1 + e^{v\omega})^2} + \frac{T_w^T}{s}u - \vartheta. \tag{4.42}$$

Since $y/(y + 1)^2 \le 1/4$ for any $y > 0$ (it can be easily proved based on the observation that the left-hand side of this inequality reaches the maximum when $y = 1$), we have the following inequality:

$$\frac{v e^{v\omega}}{(1 + e^{v\omega})^2} + \frac{T_w^T}{s}u - \vartheta \le \frac{v}{4} - (\vartheta - T_w^T u/s). \tag{4.43}$$

If the right-hand side of (4.43) is negative, we have $\frac{\partial U^p}{\partial \omega} < 0$, and $\omega = 0$ yields the maximum utility for the PO. If $v/4 - (\vartheta - \mathcal{T}_w^T u/s) \geq 0$, with the increase of ω, U^p would first decrease, then increase, and finally decrease again. In order to find the point of ω which results in the maximum U^p, let $\frac{\partial U^p}{\partial \omega} = 0$ and calculate ω as

$$\omega = \frac{1}{v} \ln \left(\frac{v - 2\left(\vartheta - \mathcal{T}_w^T u/s\right) \pm \sqrt{v^2 - 4v\left(\vartheta - \mathcal{T}_w^T u/s\right)}}{2(\vartheta - \mathcal{T}_w^T u/s)} \right).$$

According to the trend of function U^p, it is not difficult to figure out that, if $\omega_1 < \omega_2$, the maximum value of U^p would be achieved at ω_2. Hence, the best response of the PO can be finally obtained as

$$\omega^* = \begin{cases} \dfrac{1}{v} \ln \left(\dfrac{v - 2\left(\vartheta - \frac{\mathcal{T}_w^T u}{s}\right) + \sqrt{v^2 - 4v\left(\vartheta - \frac{\mathcal{T}_w^T u}{s}\right)}}{2(\vartheta - \frac{\mathcal{T}_w^T u}{s})} \right), & \text{if } \dfrac{v}{4} - (\vartheta - \dfrac{\mathcal{T}_w^T u}{s}) \geq 0; \\[4mm] 0, & \text{otherwise.} \end{cases}$$

The NE of the pricing game $\mathcal{G}_{su}$ will be then updated by substituting ω^* into (4.35). At the end, each PO i would recall $\omega_i^* \mathcal{T}_w^T$ spectrum in order to maximize its total utility and each winning SU $j, j \in W_i$, would decrease its payment to $(1 - \beta_j^*)p_j$ so as to offset its risk from spectrum recall.

4.5 Performance Analyses

In this section, the economic robustness of the overall TAGS mechanism is verified, and then the detailed time-line of TAGS is presented.

4.5.1 Proof of Economic Properties

Since the TAGS mechanism consists of two sequential stages, we need to re-examine that whether the overall mechanism can produce an economically feasible solution.

Theorem 4.3. *The TAGS mechanism is incentive-compatible.*

Proof. For each SU, its strategies are its bidding price in the first stage and its payment reduction in the second stage. Lemma 4.2 has proved that truthful bidding is a dominant strategy for each SU in the first stage. In addition, the decision process of payment reduction in the second stage is indeed a complete information game where incentive compatibility is not a concern. Hence, all SUs will be truthful throughout the TAGS mechanism.

For each PO, the quantity of spectrum recall is determined by the NE of Stackelberg game in the second stage. Thus, we only need to examine whether each PO would like to auction all its idle spectrum in the first stage. Since each PO i is enabled to recall spectrum to satisfy the potentially increasing demand from its own PUs, it takes no risk on balancing the amount of spectrum for leasing and reservation. Therefore, the amount of auctioned spectrum C_i, which results in a maximum auction revenue R_i, can also lead to a maximum utility for PO i, i.e.

$$\arg\max_{C_i} U_i^p = \arg\max_{C_i} R_i, \quad \forall i \in \mathcal{M}. \tag{4.44}$$

Since the auction revenue R_i is obviously a non-decreasing function of C_i, in order to maximize U_i^p, PO i should auction all its idle spectrum in the first stage. Therefore, both buyers and sellers will bid truthfully in the TAGS mechanism. $\square$

Theorem 4.4. *The TAGS mechanism is individual-rational for POs.*

Proof. Consider the expression of U^p in (4.39). In the first stage, each PO makes the contract with the *broker* and accordingly, the *broker* calculates the payoff R_i for each PO i. However, POs may break the contract by increasing ω if and only if their utilities could be enhanced. Since POs are leaders in the Stackelberg game, which means that they are aware of all the information, we have

$$U_i^p \geq R_i \geq 0, \quad \forall i \in \mathcal{M}. \tag{4.45}$$

Hence, the utilities of all POs in TAGS must be larger than or equal to 0. $\square$

Theorem 4.5. *The TAGS mechanism is individual-rational for all winning SUs in W_i when the spectrum recall ratio ω_i is no more than a certain threshold Γ_i.*

Proof. First, let us consider the case when $\omega_i = 0$. In this case, the utility of each SU j is only determined by the spectrum auction, which has already been proved to be non-negative in Lemma 4.1.

Now, consider the utility function of SUs in (4.20) when $\omega_i \neq 0$. In order to ensure the utility of SU j to be non-negative, the sum of the compensation and payment reduction should be greater than or equal to the utility loss caused by QoS degradation, i.e.,

$$(1 - \beta_j)p_j \geq (\sigma_j - \kappa_j)r_j \log_2(1 + \eta_j), \quad \forall j \in W_i. \tag{4.46}$$

By substituting β_j at the NE point in (4.37) and the recall distribution r_j in (4.14), we have

$$(1 - \Pi_j\omega_i)p_j \geq (\sigma_j - \kappa_j)\omega_i T_w^T \frac{\frac{D_j}{\Pi_j p_j}}{\sum \frac{D_k}{\Pi_k p_k}} \log_2(1 + \eta_j). \tag{4.47}$$

Let $J_{4,j} = (\sigma_j - \kappa_j)\mathcal{T}_w^T \frac{\frac{D_j}{\Pi_j p_j}}{\sum \frac{D_k}{\Pi_k p_k}} \log_2(1 + \eta_j)$. The above inequality finally yields

$$\omega_i \le \frac{p_j}{J_{4,j} + \Pi_j p_j}, \quad \forall j \in W_i. \tag{4.48}$$

Therefore, we can draw a conclusion that when $\omega_i \le \Gamma_i = \min\{\frac{p_j}{J_{4,j} + \Pi_j p_j}, \forall j \in W_i\}$, all SUs in W_i could be guaranteed with non-negative utilities.   □

Theorem 4.5 implies that the maximum amount of recalled spectrum declared by each PO i should be $\min\{\omega_i^*, \Gamma_i\}$ so as to maintain the rationality of the mechanism while maximizing its own utility.

4.5.2   Detailed Procedure of the Two-Stage Sharing Mechanism

The procedure of TAGS is summarized as follows:

- At the beginning of each frame, all POs and SUs report their bids to the *spectrum broker*.
- The *spectrum broker* collects the received sealed information and then builds up a combinatorial spectrum auction. The winner determination and payment scheme follow the approximation mechanism presented in Sect. 4.3.
- After the auction, each PO informs its own winning SUs a maximum quantity of recalled spectrum and along with its recall scheme as in (4.14). Each SU then decides a payment reduction to offset the risk, while at the same time trying to avoid large utility degradation. POs in turn derive the spectrum recall ratio distributed on each winning SU when the payment strategies have been finally determined.
- In the rest time of each frame, POs would recall spectrum to meet the increase of their own spectrum demand for newly arrived PUs. The total amount of spectrum recalled by each PO and the distributed recall ratio on each winner follow the strategies made in the previous step. At the end of each frame, POs compensate each winner based on (4.17).

4.6   Numerical Results

In this section, simulations are conducted to evaluate the TAGS mechanism. The performance of two stages is first illustrated separately and then the overall performance of TAGS in terms of spectrum utilization and utilities is presented.

Consider a CR network with m POs and n SUs randomly scattered in a 200×200 geographic area, where m varies from 5 to 50 and $n = 100$. Each PO i owns a total spectrum bandwidth $\mathcal{T}_w^i$ randomly in $[20, 100]$ MHz and a same auction radius $R = 80$. Each primary user has the same demand s which is selected as 1, 2 or 3 MHz. The activities of PUs among different POs are heterogeneous with an arrival rate $\lambda_i = 1, 2$, or 3 and a service rate μ_i randomly selected from 0.1 to 0.2. For simplicity, assume that all PUs have the same waiting time requirement $\gamma = 6.25 \times 10^{-4}$ s and the same individual utility $u = 2 \times 10^5$. The spectrum demand D_j and the SNR η_j of each SU j are randomly selected in $[1, 10]$ MHz and $[100, 200]$, respectively. Furthermore, the weight indices are defined as follows: $\psi_i = [1, 2, 3]$ for each PO i; $\sigma_j = 1$ and κ_j is randomly chosen in $[0, 1]$ for each SU j. Suppose that the length of each frame $T = 6$ s and we observe all results from $150T$ to $200T$. Note that some parameters may vary according to evaluation scenarios.

Figure 4.3 shows the social welfare (averaged from $150T$ to $200T$) achieved by different algorithms in the first-stage spectrum auction. For comparison purpose, the pure allocation (PA) is simulated as the benchmark, which iteratively selects the unallocated SU with largest spectrum demand and assigns it to POs regardless of its bidding price. It shows that the introduced WDP algorithm can achieve much higher social welfare than PA, and this superiority becomes more obvious when the number of POs increases. The reason is that the larger number of POs enhances the probability for better spectrum allocation. Moreover, by comparing between the initial solution of WDP algorithm and the improved one, we can clearly observe that the improvement processes can effectively increase the social welfare. By further considering the polynomial computational complexity, we can conclude that the WDP algorithm is feasible for practical implementations.

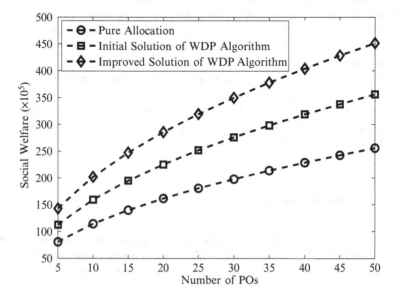

Fig. 4.3 The performance of the WDP algorithm

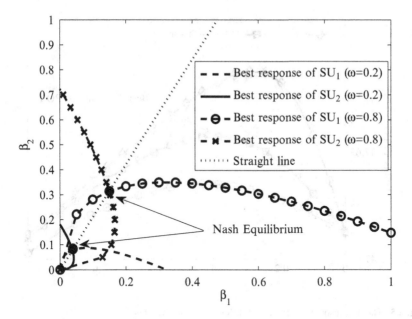

Fig. 4.4 The existence and uniqueness of NE in $\mathcal{G}$

In Fig. 4.4, the properties of the NE in $\mathcal{G}_{su}$ are examined. For concise and clear demonstration, the situation with $t = 2$ is considered. Figure 4.4 shows that for any value of recall ratio ω, there is a unique intersection point between the best responses of SU 1 and SU 2. Note that according to the analysis in Sect. 4.4.2.1, $\beta_1 \neq 0$ and $\beta_2 \neq 0$. Otherwise, (4.14) becomes meaningless. In addition, the points of NE under different recall ratios locate on a straight line, which verify the calculation in (4.36) that the ratio between β_1^* and β_2^* is always a constant, i.e., $\beta_2^*/\beta_1^* = \mathcal{B}_1/\mathcal{B}_2$.

Figure 4.5 reveals a PO's utility with the increase of its spectrum recall ratio. Given the NE of $\mathcal{G}_{su}$, the PO can take initiative by deciding the optimal value of ω so as to maximize its utility U^p. The trend of these curves indicates that U^p first increases with ω. That is because with the increase of ω, more spectrum is available for recall so that more utility from PUs and less penalty for QoS degradation are achieved. However, after a certain point, since the compensation to winning SUs becomes dominant, U^p decreases. The ω which results in the highest utility is the best response. We can also observe from the figure that a larger value of penalty weight ψ results in a larger value of the best response for PO. This is because the spectrum requirement from PUs increases when the penalty becomes larger. As a consequence, PO would prefer to recall more spectrum to satisfy its own users' demands.

Figure 4.6 compares the spectrum utilization ratio (i.e., the ratio between the amount of occupied spectrum and the total amount of POs' spectrum) with and without the TAGS mechanism. When TAGS is not applied, both spectrum auction and recall pricing game are not activated so that such ratio is only determined by

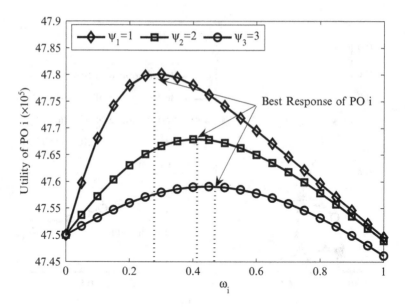

Fig. 4.5 The best response of PO i with different penalty weights

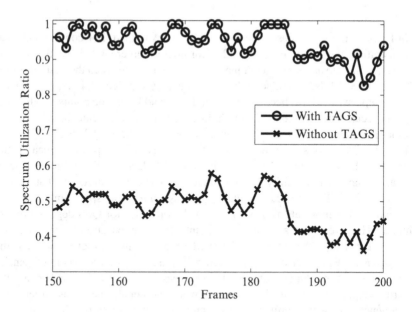

Fig. 4.6 The comparison of spectrum utilization with and without TAGS

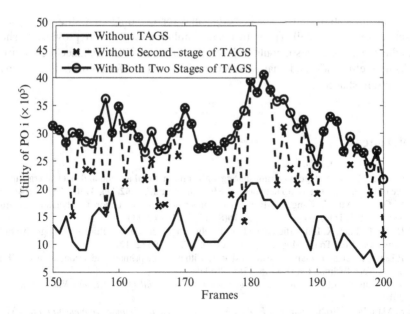

Fig. 4.7 The comparison of PO's utility with and without two stages

the spectrum usage of PUs. Nearly half of the spectrum is under-utilized in this circumstance though the spectrum demands from PUs can be fully satisfied. On the contrary, with the employment of TAGS, a balance on spectrum utilization for both SUs and PUs can be reached so that a much higher utilization ratio is achieved.

Figure 4.7 shows the superiority of TAGS on the utility of a randomly selected PO. Without TAGS, the PO's utility, U^p, only includes its own PUs' revenue. According to the analysis of Fig. 4.6, since plenty of spectrum is under-utilized without running spectrum sharing mechanism, U^p is relatively low. It can also be seen from the figure that the PO has highly fluctuating utility without the second stage of TAGS. This is because even though more utility is produced from the first-stage auction, the increasing demand from its own PUs leads to a large penalty on U^p due to the QoS degradation. In summary, the curve with both stages of TAGS apparently shows the best performance since spectrum recall is enabled and the best recall quantity is determined from the Stackelberg game in the second stage.

4.7 Summary

In this chapter, the spectrum sharing issue among multiple heterogeneous POs and SUs in recall-based CR networks has been discussed. To address the complexity of this system model, a two-stage mechanism, called TAGS, was introduced, which consisted of a geographically restricted combinatorial spectrum auction for initial spectrum allocation and a Stackelberg game for deciding best strategies toward potential spectrum recall. Theoretical and simulation results were provided to show that the TAGS mechanism could improve the utilities of POs, enhance the spectrum utilization efficiency, and guarantee economic incentives for all users to participate in spectrum sharing.

References

1. C. Yi, J. Cai, Two-stage spectrum sharing with combinatorial auction and stackelberg game in recall-based cognitive radio networks. IEEE Trans. Commun. **62**(11), 3740–3752 (2014)
2. L. Gao, Y. Xu, X. Wang, MAP: multiauctioneer progressive auction for dynamic spectrum access. IEEE Trans. Mobile Comput. **10**(8), 1144–1161 (2011)
3. D. Xu, E. Jung, X. Liu, Efficient and fair bandwidth allocation in multichannel cognitive radio networks. IEEE Trans. Mobile Comput. **11**(8), 1372–1385 (2012)
4. P. Lin, J. Jia et al., Dynamic spectrum sharing with multiple primary and secondary users. IEEE Trans. Veh. Technol. **60**(4), 1756–1765 (2011)
5. L.A. Wolsey, G.L. Nemhauser, *Integer and Combinatorial Optimization* (Wiley, New York, 2014)
6. S. Martello, P. Toth, *Knapsack Problems: Algorithms and Computer Implementations* (Wiley, New York, 1990)
7. C. Yi, J. Cai, Combinatorial spectrum auction with multiple heterogeneous sellers in cognitive radio networks, in *Proceedings of the IEEE ICC*, June 2014, pp. 1626–1631
8. R.B. Myerson, *Game Theory: Analysis of Conflict* (Harvard University, Cambridge, 1991)
9. R. Yates, A framework for uplink power control in cellular radio systems. IEEE J. Sel. Areas Commun. **13**(7), 1341–1347 (1995)
10. S. Jayaweera, T. Li, Dynamic spectrum leasing in cognitive radio networks via primary–secondary user power control games. IEEE Trans. Wireless Commun. **8**(6), 3300–3310 (2009)
11. H. Lin, M. Chatterjee, S. Das, K. Basu, ARC: an integrated admission and rate control framework for competitive wireless CDMA data networks using noncooperative games. IEEE Trans. Mobile Comput. **4**(3), 243–258 (2005)
12. M. Xiao, N. Shroff, E. Chong, Utility-based power control in cellular wireless systems, in *Proceedings of the IEEE INFOCOM*, 2001, pp. 412–421
13. J. Zhang, Q. Zhang, Stackelberg game for utility-based cooperative cognitive radio networks, in *Proceedings of the ACM MobiHoc*, 2009, pp. 23–31

Chapter 5
Online Spectrum Allocation Mechanism

5.1 Introduction

Without the synchronization of allocation periods, it is impossible to run the spectrum sharing mechanism in an offline manner. Thus, in this chapter, we investigate the online spectrum allocation problem in CR networks with uncertain activities of both PUs and SUs. In this system model, there is a PBS who owns multiple licensed radio channels and is responsible to protect PUs' spectrum usages. At the same time, the PBS also runs an online auction to lease its idle channels to SUs who request and access spectrum on the fly. By considering a more practical situation that the PBS has no a priori information of PUs' activities,[1] the PBS may suffer a great penalty if it is only eager to improve its potential auction revenue while ignoring its own PUs' spectrum usages. On the other hand, if the PBS reserves channels excessively to completely protect its own PUs, it may lose economic profits from the spectrum auction. To balance the penalties introduced by incomplete services for PUs and the auction profits from granted SUs' spectrum requests, we present a new approach, called *virtual online double spectrum auction (VIOLET)* mechanism [1]. In this mechanism, the concept of virtual spectrum sellers is introduced to describe the channel uncertainties. The well-designed online admission and pricing scheme of VIOLET can ensure non-deficit utility of the PBS while resisting mendacious behaviors from selfish SUs. Theoretical analyses prove that the VIOLET mechanism is economic robust in terms of budget-balance, individual rationality, and incentive compatibility. In addition, simulation results show that the VIOLET mechanism can improve the utility of the PBS, enhance spectrum utilization, and achieve better satisfaction of SUs.

---

[1]Note that this model is different from the one with known statistics of PUs' activities in either Chap. 3 or 4.

© The Author(s) 2016
C. Yi, J. Cai, *Market-Driven Spectrum Sharing in Cognitive Radio*,
SpringerBriefs in Electrical and Computer Engineering,
DOI 10.1007/978-3-319-29691-3_5

The rest of this chapter is organized as follows: Sect. 5.2 illustrates the system model. Section 5.3 describes the VIOLET mechanism in detail with its design goals, admission, and pricing scheme. Section 5.4 provides solid theoretical analyses for economic properties. Simulation results are presented in Sect. 5.5. Finally, Sect. 5.6 concludes this chapter.

5.2   System Model

Consider a CR network with one PBS and m subscribed PUs registered with it. Each PU demands exclusive-usage of one licensed channel so that the PBS owns the license of m homogeneous orthogonal spectrum channels and is responsible for serving all its PUs' communication requests. Obviously, the PBS can serve all its registered PUs simultaneously, and there is no competition among PUs. However, the activities of PUs are uncertain, which means that any PU may declare its spectrum request at any time for any long period. Since PUs have already signed a service contract with the PBS, they cannot suffer any delay. Moreover, if a request from PU j is not served, the PBS will be punished by a predetermined penalty ϕ_j. In CR networks, PUs have higher channel access priority than SUs, and their services are protected due to the contracts pre-signed with the PBS. Thus, it is reasonable to assume that information provided by PUs are always truthful.

The network also includes some SUs who are willing to buy the usage of idle spectrum at any time. Assume that SUs $\mathcal{S} = \{su_1, su_2, \ldots\}$ will request spectrum usage on the fly. Since in wireless networks, SUs may be located in different geometric areas, this location-dependent feature allows potential spatial reuse of spectrum among time overlapped SUs' requests. To capture such spatial reuse, in this chapter, the SU network is modeled as a conflict graph $\mathcal{H} = (\mathcal{S}, E)$ by applying existing methods [2, 3], where $\mathcal{S}$ is the vertex set corresponding to the requests of SUs and two SUs, su_i and su_k, form an edge $(su_i, su_k) \in E$ if and only if they cannot access the same channel simultaneously. We further limit our discussion on the scenario that each SU only demands a single time-frequency chunk in each request, i.e., a single time slot from one channel. Such scenario has been widely applied for file transmissions (such as HTTP/FTP) [4], where discrete radio resources are valuable. Let $e_1, e_2, \ldots, e_i, \ldots$, be the sequence of requests over a long time period $T = [0, T]$ which consists of T normalized time slots. Each request is expressed as $e_i = (su_i, a_i, d_i, v_i)$, called the profile of request from SU su_i, which indicates that SU su_i declares a spectrum request at arrival time $a_i \in T$ for a service within $[a_i, d_i]$, where $d_i \in T$ is the deadline for fulfilling this request and $v_i \in \mathbb{R}^+$ represents its valuation for receiving the service. If $d_i - a_i > 1$, some delay tolerance exists for the service of e_i. Note that each SU may request multiple non-overlapped spectrum usages. However, since there is no need to differentiate SUs if their conflict graph is already known, for notation simplicity, we omit index su_i in the following context and use $e_i = (a_i, d_i, v_i)$ to denote the i-th request from SUs.

Let the PBS be the spectrum auctioneer, and its objective be to improve spectrum utilization and its own overall utility.

5.3   Virtual Online Double Spectrum Allocation Mechanism

In this section, we first introduce the idea of virtual sellers and explain how the channel uncertainties can be represented by the online activities of virtual sellers. Then, the admission and pricing scheme of VIOLET mechanism are described.

5.3.1   Virtual Spectrum Sellers

In order for the PBS to make allocation decisions for SUs while ensuring itself a non-deficit utility, a novel concept of virtual spectrum sellers is used in the auction based on the uncertain activities of PUs. Specifically, the PBS creates $I(t)$ virtual sellers at any time instant $t \in T$, where $I(t)$ equals the number of idle channels at the end of $t - 1$. The generation of virtual sellers is as follows.

1. For PU request arriving at t, the PBS generates a corresponding virtual seller j with $c_j = \{(a_j = t, d_j = t + 1, v_j = \phi_j)\}$, which indicates that the seller who arrives at a_j has no patience (departs in one time slot) and wants to sell its resource at an asking price of ϕ_j. Note that setting $d_j = a_j + 1$ only means that the PU's request has no delay tolerance. In fact, the PU can request multiple time slots. Once it is granted, it will stay in the system till its requested service is finished.
2. If the number of newly arrived PUs, $A(t)$, is less than $I(t)$, the PBS automatically adds $I(t) - A(t)$ virtual sellers, all with asking price equal to 0, i.e., $c_j = \{(a_j = t, d_j = t + 1, v_j = 0)\}$.

Note that, unlike traditional auctions, sellers in this scenario are not real, which means that they do not have real incomes. In other words, if a virtual seller wins the auction, its income (which is actually the penalty on the PBS) will be exactly its asking price.

5.3.2   Design Requirements

Let e^t and c^t denote the set of requests from SUs (buyers) and virtual sellers in t, respectively. Further define $e = (e^1, \ldots, e^t, \ldots, e^T)$ and $c = (c^1, \ldots, c^t, \ldots, c^T)$ denote the complete request profile over T. Since SUs are self-interested and their requests are private information, SU i may misreport its request, i.e., $\hat{e}_i = (\hat{a}_i, \hat{d}_i, \hat{v}_i) \neq e_i$, if it could benefit from such behavior. Similar as most of the online mechanisms [4, 5], we assume that there is no misreport of either early-arrival or later-departure in the system. In practice, reporting $\hat{a}_i < a_i$ or $\hat{d}_i > d_i$ may probably lead to a service beyond the range of $[a_i, d_i]$, which is not expected by the SU. In addition, all SUs' requests are assumed to have a bounded patience, i.e., $d_i \leq a_i + \Delta$ and $\Delta \neq \infty$.

For bidder (both buyers and sellers) profile $B = e \cup c$, let B^{t-} represent the profile with arrival time no later than t. An online double spectrum auction mechanism, $\mathcal{M}_c \triangleq (x, p)$, defines an allocation decision $x = \{x^t\}^{t \in T}$ and a payment scheme $p = \{p^t\}^{t \in T}$, where x^t and p^t denote the allocation and payment vector at each time slot t, respectively. For each bidder k with its profile $B_k \in B$, further define $x_k^t(B^{t-}) \in \{0, 1\}$ to indicate whether bidder k wins no later than t and $p_k^t(B^{t-}) \in \mathbb{R}$ to indicate its payment. Note that $p_k^t(B^{t-}) \geq 0$ if bidder k is a buyer, i.e., $B_k \in e^{t-}$, and $p_k^t(B^{t-}) \leq 0$ if bidder k is a seller, i.e., $B_k \in c^{t-}$. For a feasible mechanism, it must satisfy the condition that $x_k^t(B^{t-}) = 1$ in at most one slot $t \in [a_k, d_k]$ and zero for other values of t.

Now, the corresponding economic properties in terms of *budget-balance*, *individual rationality* and *incentive compatibility* can be defined as follows.

Definition 5.1 (Budget-Balance). $\mathcal{M}_c \triangleq (x, p)$ is budget-balanced if the utility of the PBS is always non-deficit, i.e.,

$$U_{pbs}^t = \sum_{B_k \in B^{t-}} \sum_{t' \in [a_k, \min(t, d_k)]} p_k^{t'}(B^{t'-}) \geq 0, \quad \forall t \in T.$$

This property ensures that the PBS can always benefit from running the auction, even though it suffers potential penalties from its own PUs.

Since individual rationality and incentive compatibility are only required for SUs, let $x_i(e) = \sum_{t' \in [a_i, d_i]} x_i^{t'}(e^{t'-})$ and $p_i(e)$ indicate whether e_i wins or not and its payment, respectively.

Definition 5.2 (Individual Rationality). $\mathcal{M}_c \triangleq (x, p)$ is individual-rational for all SUs' requests, if no e_i pays more than its valuation, i.e.,

$$v_i(x(e)) - p_i(e) \geq 0, \quad \forall i \in \{i | B_i \in e\},$$

where $v_i(x(e))$ represents the valuation of e_i given the allocation $x(e)$, i.e., $v_i(x(e)) = v_i$ if e_i wins in $x(e)$, and $v_i(x(e)) = 0$, otherwise. With this property, the utility of SUs can be always non-negative which provides them incentives to participate.

Before introducing the definition of truthfulness, we define $e_{-i} \in e$ as the set of other requests except e_i, and $\Omega(e_i)$ as the set of potential misreports of e_i.

Definition 5.3 (Incentive Compatibility). $\mathcal{M}_c \triangleq (x, p)$ is incentive-compatible or truthful if no SU can improve its utility by misreporting its type, i.e.,

$$v_i(x(e_i, e_{-i})) - p_i(e_i, e_{-i}) \geq v_i(x(\hat{e}_i, e_{-i})) - p_i(\hat{e}_i, e_{-i}), \quad \forall \hat{e}_i \in \Omega(e_i), \forall i \in \{i | B_i \in e\}.$$

This property is essential for a robust auction mechanism. It resists market manipulation and ensures auction efficiency and fairness.

5.3.3 Admission and Pricing Mechanism

Here, the design of the VIOLET mechanism is studied with the consideration of the reusability of wireless spectrum and the satisfaction of all required economic properties.

5.3.3.1 Grouping SUs' Requests

At any time instant t, the PBS can group the outstanding SUs' spectrum requests based on the pre-determined conflict graph $\mathcal{H}$. For SUs that do not interfere with each other, their requests are grouped into the same group and each of them can be assigned the same spectrum chunks. Such process is equivalent to finding independent sets of the conflict graph and is processed privately by the PBS. Specifically, the PBS can recursively select a node in current conflict graph and include it to the set, eliminate the chosen node and its neighbors, and update the topology of the remaining nodes.

Let $\xi_1^t, \xi_2^t, \dots, \xi_{\mathcal{L}(t)}^t$ denote the $\mathcal{L}(t)$ buyers' groups formed at t. Each ξ_l^t is regarded as a super buyer with $|\xi_l^t|$ non-conflict members. Then, the group bid g_l^t can be calculated as

$$g_l^t = \min\{\hat{v}_i | e_i^t \in \xi_l^t\} \times |\xi_l^t|, \tag{5.1}$$

where $\min\{\hat{v}_i | e_i^t \in \xi_l^t\}$ represents the minimum reporting valuation of a request in group ξ_l^t.

5.3.3.2 Myopic Matching and Pricing

For a specific time slot t, we have $I(t)$ virtual sellers. For clarity, let $q_j^t \in \{v_j | B_j \in c^t\}$ denote the asking price of virtual seller j at t. Following the static McAfee [6] matching rule, the PBS sorts all buyers' group bids and sellers' asking prices collected at t in a non-increasing and a non-decreasing order, respectively, i.e.,

$$g_1^t \geq g_2^t \geq \dots \geq g_{\mathcal{L}(t)}^t;$$
$$q_1^t \leq q_2^t \leq \dots \leq q_{I(t)}^t.$$

Now, match the above two orders one by one, and let r index the last profitable pair, i.e.,

$$r = \underset{r' \leq \min\{\mathcal{L}(t), I(t)\}}{\arg\max} \; g_{r'}^t - q_{r'}^t \geq 0. \tag{5.2}$$

Then, the first $r - 1$ buyer groups will win the auction and the first $r - 1$ virtual sellers will be traded at t. On the other hand, the rest of virtual sellers will lose and their corresponding number of channels will be reserved for PUs. To guarantee myopic truthfulness, each winning buyer group $\boldsymbol{\xi}_w^t$ will be charged by the r-th buyer group's bid g_r^t (the highest losing bid), and such group payment is shared equally among all SUs' requests in group $\boldsymbol{\xi}_w^t$, i.e.,

$$p_i^t = g_r^t / |\boldsymbol{\xi}_w^t|, \quad \forall e_i \in \boldsymbol{\xi}_w^t. \tag{5.3}$$

Any losing SUs' request does not need to pay and no virtual seller is paid with real profit. However, the PBS would be penalized for trading virtual sellers with their asking prices. Therefore, the myopic utility of the PBS can be expressed as

$$U_{myopic}^t = \sum_{j=1}^{r-1} (g_r^t - q_j^t). \tag{5.4}$$

5.3.3.3 Online Payment Calculation

In order to resist both bid- and time-based cheating from SUs while maintaining budget balance for the PBS, a novel online pricing scheme is applied for each SU's request e_i:

- *Upon arrival*: Consider the myopic double auction in all its possible early arrival time $t' \in [\hat{d}_i - \Delta, \hat{a}_i - 1]$ with its reported bid, where Δ is the defined bounded patience. If it would lose in all its early arrival times, we set its admission price $\mathcal{P}^m(\hat{a}_i, \hat{d}_i, e_{-i}) = 0$. Otherwise,

$$\mathcal{P}^m(\hat{a}_i, \hat{d}_i, e_{-i}) = \max\{p_i^{t'} | t' \in [\hat{d}_i - \Delta, \hat{a}_i - 1]\}, \tag{5.5}$$

where $p_i^{t'}$ indicates the myopic pricing that such request has to pay at t'.
- *During active period*: For any t from $\hat{a}_i$ to $\hat{d}_i$, if the request wins the myopic auction at t, it will be selected as a winner and its final payment is calculated as

$$p_i(\hat{a}_i, \hat{d}_i, e_{-i}) = \max(p_i^t, \mathcal{P}^m(\hat{a}_i, \hat{d}_i, e_{-i})). \tag{5.6}$$

If the request cannot win in any time $t \in [\hat{a}_i, \hat{d}_i]$, its payment is set as 0.

Corollary 5.1. *The double spectrum auction mechanism, i.e., VIOLET, can be reduced to a sequence of myopic TRUST mechanism [7] if all SUs' spectrum requests are impatient to any delay, i.e., $\Delta = 1$.*

5.4 Proof of Economic Properties

In this section, we prove that the VIOLET mechanism satisfies all desired economic properties through theoretical analyses.

5.4.1   Budge-Balance and Individual Rationality

Theorem 5.1. *The VIOLET mechanism is budget-balanced for the PBS and individually rational for all the SUs' requests.*

Proof. According to the myopic matching rule, we can simply observe that only the bid-ask pair with bid greater than ask would be selected to trade at a time t. Moreover, since the PBS does not need to pay for winning virtual sellers and its penalty equals the sum of exactly their asking prices, the myopic utility of the PBS, $U^t_{myopic} \geq 0$. In addition, the final payment of each winning SU's request is calculated as the maximum payment it has made if it could win in any time instant from its possible earliest arrival to its winning instant. Therefore, the utility of the PBS at any time t is even larger than its myopic utility, i.e.,

$$U^t_{pbs} \geq U^t_{myopic} \geq 0, \quad \forall t \in T. \tag{5.7}$$

Thus, the VIOLET mechanism ensures budget-balance for the PBS.

Individual rationality can be immediately proved from the pricing scheme. First, an SU has to pay only if its request e_i could win at $t \in [a_i, d_i]$. Even though the payment $p_i(e)$ is not simply equal to the market clearing price at t, such payment must be calculated from one of the myopic double auction during $[d_i - \Delta, t]$, in which e_i could win. Based on the myopic pricing rule, each buyer group declares its group bid based on the minimum individual valuation of its members and if it wins the auction, it pays the highest bid from losing groups. Thus,

$$v_i \geq p_i(e) = p^\tau_i, \quad \exists \tau \in [d_i - \Delta, t]. \tag{5.8}$$

Hence, the VIOLET mechanism guarantees individual rationality.   □

5.4.2   Incentive Compatibility

In order to prove the incentive compatibility of the VIOLET mechanism for all SUs' requests, the following prerequisite technical lemmas are first examined.

Lemma 5.1. *At time slot $t \in [a_i, d_i]$, if an SU's request e_i wins in the VIOLET, then for fixed a_i, d_i and e_{-i}, the myopic payment, p^t_i, is independent of its bid value v_i.*

Proof. Proving the myopic value-independency is equivalent to proving that if e_i wins at t by bidding v_i or $\hat{v}_i$, the myopic payment charged from e_i is the same for both cases. Let ξ^t_l denote the group that e_i belongs to, and g^t_l ($\hat{g}^t_l$) be the bids of ξ^t_l when e_i bids v_i ($\hat{v}_i$). If e_i can win in both groups, the market clearing price of ξ^t_l is determined by the same highest losing group bid, g^t_r, ranked after g^t_l and $\hat{g}^t_l$. Furthermore, the size $|\xi^t_l|$ also keeps the same for bidding v_i or $\hat{v}_i$ from e_i. Therefore, our claim holds since $p^t_i = g^t_r / |\xi^t_l|$.   □

Lemma 5.2. *No matter the SU's request e_i wins or loses in VIOLET, for fixed a_i, d_i and e_{-i}, reporting $\hat{v}_i > v_i$ for which it can win the auction will lead to a payment, $\hat{p}_i(e)$, larger or equal to $p_i(e)$. More seriously, such misreporting may also break the property of individual rationality.*

Proof. Two cases have to be discussed, i.e., whether or not e_i can win with truthful value v_i.

C1:   If e_i wins by bidding v_i, misreporting $\hat{v}_i \geq v_i$ would not change the myopic payment p_i^t at its trading time t (Lemma 5.1). However, since $\hat{v}_i$ may result in more winnings during $[d_i - \Delta, a_i - 1]$, the admission price $\mathcal{P}^m(a_i, d_i, e_{-i})$ of e_i would monotonically increase according to (5.5). With the calculation of final payment in (5.6), we can directly observe that $\hat{p}_i(e) \geq p_i(e)$.

C2:   If e_i loses by bidding v_i, it also means that e_i cannot win any myopic double auction during $[a_i, d_i]$. Suppose that $\exists t' \in [a_i, d_i]$ when e_i could win with $\hat{v}_i$. Let $g_l^{t'}$ and $\hat{g}_l^{t'}$ be the bids of e_i's group $\xi_l^{t'}$ when e_i reports v_i and $\hat{v}_i$ at t', respectively. Since the auction results of e_i are changed by bidding $\hat{v}_i > v_i$, the bid of e_i must be the lowest one in $\xi_l^{t'}$, i.e., $v_i = g_l^{t'}/|\xi_l^{t'}|$. Moreover, since e_i loses by bidding v_i while winning by bidding $\hat{v}_i$, we have $g_l^{t'} \leq g_p^{t'} \leq \hat{g}_l^{t'}$, where $g_p^{t'}$ is the market clearing price at t'. Therefore, the utility of e_i with bid $\hat{v}_i$ is $(v_i - g_p^{t'}/|\xi_l^{t'}|) \leq 0$.

Consequently, with *C1* and *C2*, Lemma 5.2 is proved.   □

With Lemmas 5.1 and 5.2, we can further prove that VIOLET mechanism satisfies the price-based characterization [8], which is essential for establishing incentive compatibility in online auctions.

Definition 5.4 (Price-Based Characterization [8]). An online auction is price-based if there is a value-independent pricing scheme such that e_i wins if and only if $p_i(a_i, d_i, e_{-i}) \leq v_i$ and the payment of a winning e_i, $p_i(e) = p_i(a_i, d_i, e_{-i})$.

Lemma 5.3. *The VIOLET is a price-based mechanism.*

Proof. From Lemma 5.1, we know that if an SU's request e_i wins at a time $t \in [a_i, d_i]$, then for fixed a_i, d_i and e_{-i}, its myopic payment p_i^t is independent of its required value v_i. Moreover, we know that such payment must be less than or equal to the minimum bid value across all SUs' requests that will be granted at t. From Lemma 5.2, we also see that such payment must be greater than all bid values from losing requests. In conclusion, there is a value-independent price for trading at t, which is the minimal value that a winning SU's request could have bid at that time instant. Finally, note that for fixed $[a_i, d_i]$ and fixed e_i, the request e_i would always win at the same time with its bid value greater than the myopic payment. Hence, $p_i(e) = p_i(a_i, d_i, e_{-i})$ as required and Lemma 5.3 holds.   □

Furthermore, the next lemma prove that no SU's request could benefit from misreporting $\hat{a}_i > a_i$ or $\hat{d}_i < d_i$.

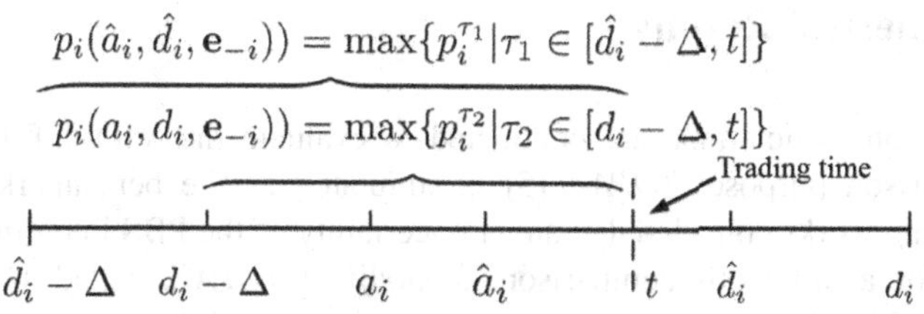

$$\underbrace{p_i(\hat{a}_i, \hat{d}_i, \mathbf{e}_{-i})) = \max\{p_i^{\tau_1} | \tau_1 \in [\hat{d}_i - \Delta, t]\}}$$

$$p_i(a_i, d_i, \mathbf{e}_{-i})) = \max\{p_i^{\tau_2} | \tau_2 \in [d_i - \Delta, t]\}$$

Fig. 5.1 Impact of time cheating on the payment of a winning e_i

Lemma 5.4. *The payment of a winning e_i in VIOLET is monotonically increasing, i.e., $p_i(a_i, d_i, e_{-i}) \leq p_i(\hat{a}_i, \hat{d}_i, e_{-i})$ for $[\hat{a}_i, \hat{d}_i] \subseteq [a_i, d_i]$, and no e_i can win at an earlier time instant by reporting $[\hat{a}_i, \hat{d}_i] \subseteq [a_i, d_i]$.*

Proof. According to the online pricing scheme in (5.5) and (5.6), it can be easily derived that, for any winning e_i, we have

$$p_i(\hat{a}_i, \hat{d}_i, e_{-i}) = \max\{p_i^{t'} | t' \in [\hat{d}_i - \Delta, t]\}, \tag{5.9}$$

where t denotes its trading time.

As shown in Fig. 5.1, the payment of the winning e_i is independent of its arrival time and increased with an earlier departure since

$$\max\{p_i^{\tau_1} | \tau_1 \in [\hat{d}_i - \Delta, t]\} \geq \max\{p_i^{\tau_2} | \tau_2 \in [d_i - \Delta, t]\}, \quad \forall \hat{d}_i < d_i. \tag{5.10}$$

Thus, it only remains to prove that reporting $[\hat{a}_i, \hat{d}_i] \subseteq [a_i, d_i]$ cannot bring an earlier winning. Note that no SU's request would like to delay its winning time because the payment is obviously non-decreasing with t according to (5.9). Furthermore, reporting an earlier departure time cannot change any auction results if $t \leq \hat{d}_i$. Hence, we only need to consider the case that e_i may misreport $\hat{a}_i > a_i$. Since a later arrival has no effect on the payment due to the independency, intentionally delaying arrival will not lead to an early winning for e_i, but in turn, may possibly make it miss opportunities of early winnings. In summary, Lemma 5.4 is guaranteed. □

Taken all the above lemmas (Lemmas 5.1–5.4) together, it can be concluded that VIOLET mechanism is implemented by a price-based auction with value-independent payment scheme, and the payment of each winning e_i is monotonically increasing with (1) reporting later arrival, i.e., $\hat{a}_i > a_i$, given $\hat{d}_i = d_i$; (2) reporting earlier departure, i.e., $\hat{d}_i < d_i$, given $\hat{a}_i = a_i$; or (3) the combination of them. According to [9], the following theorem is obtained.

Theorem 5.2. *The VIOLET mechanism is incentive-compatible.*

5.5   Numerical Results

In this section, simulations are conducted to evaluate the VIOLET mechanism. For comparison purpose, *TOPAZ* [5] is simulated as the benchmark. Since no other existing works considered channel uncertainty of the PBS in online spectrum auctions, for a more fair comparison, a modified *TOPAZ*, called *M-TOPAZ*, is simulated, which adds a procedure of examining the number of available auctioned channels at the beginning of each time slot so as to fully protect PUs' services. All simulation parameters are shown in Table 5.1 and all results are based on the average over 20 runs.

Figure 5.2 shows cumulative utilities of the PBS at each time slot. Since *TOPAZ* only considers one-sided spectrum auction to SUs and ignore the potential PUs' spectrum usages, the utility of the PBS has a large variance (up to 327.4 %) due to the penalty from the uncertain activities of PUs. As a result, the property of budget-balance cannot always be guaranteed. On the other hand, the PBS suffers zero penalty with *M-TOPAZ* and obtains extra revenue from leasing its unused channels. VIOLET better balances the tradeoff between auction revenue and penalty, so that it achieves the best performance.

Figures 5.3 and 5.4 demonstrate the impact of increasing PUs' arrival rate and the number of primary channels, respectively. It is shown that the overall utility of the PBS in VIOLET decreases with the increase of PUs' arrival rate. However, its curve is descended slower than those of *TOPAZ* and *M-TOPAZ*. This is because the increase of PUs' spectrum usage results in a larger penalty on *TOPAZ* and less auction revenue on *M-TOPAZ*, respectively. In addition, Fig. 5.4 indicates that VIOLET can perform better than *M-TOPAZ* on the PBS's utility with the increase of the total number of channels.

Spectrum utilizations are compared in Fig. 5.5, which is calculated as the ratio between the total spectrum usage of winning users (both SUs and PUs) and the total

Table 5.1 Simulation settings

Parameters	Value
Area size	200×200
Conflict distance	30
Number of time slots	50
Bounded tolerance	3
Parameters	**Model**
Bid valuations	*Uniform*[0, 1]
PUs' penalties	*Uniform*[0.5, 1]
Interarrival time of PUs	*Exponential*
Service time of PUs	*Exponential*
Parameters	**[Range], Normal value**
Number of channels	[1, 10], 5
Number of SUs' requests	[100, 500], 300
PUs' arrival rate	[1, 3], 2
PUs' service rate	[0, 1], 0.5

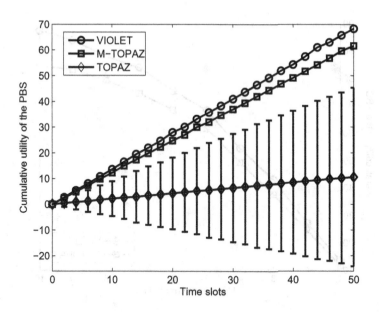

Fig. 5.2 Performance on the cumulative utility of the PBS

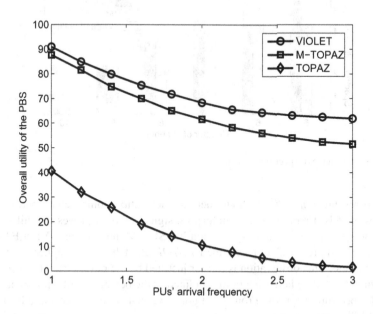

Fig. 5.3 Impact of PUs' arrival frequency on the overall utility of the PBS

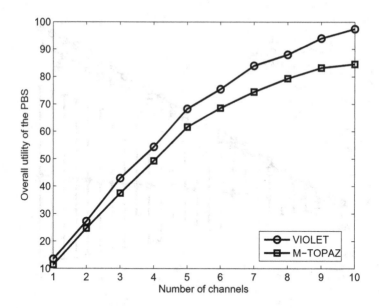

Fig. 5.4 Comparison on the overall utility of the PBS

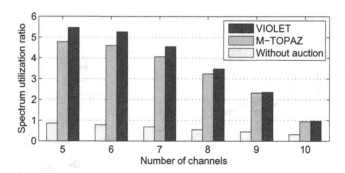

Fig. 5.5 Performance of spectrum utilization

spectrum resources $m \times T$. Without auction, such ratio is only determined by the activities of PUs. Enabling spectrum auction significantly improves the utilization due to the spectrum reuse among non-conflict SUs. Moreover, since VIOLET can accommodate more SUs' requests than *M-TOPAZ*, it has even higher spectrum utilization ratio. This observation is further justified by the buyers' satisfaction ratio as shown in Fig. 5.6, where such ratio is defined as the percentage of winners among all SUs' spectrum requests. From the figure, we can observe that the advantage of VIOLET is more and more obvious with the increase in the number of SUs' requests.

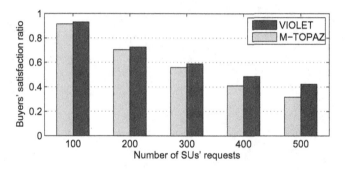

Fig. 5.6 Comparison on satisfaction ratio of SUs

5.6 Summary

In this chapter, an online spectrum allocation mechanism with uncertain activities of PUs and SUs in CR networks has been studied. In order to guarantee a non-deficit utility of the PBS so as to provide it economic incentives to participate in the auction, a virtual online double spectrum auction mechanism, called VIOLET, was introduced, in which the channel uncertainties were represented by the online activities of virtual spectrum sellers. Theoretical analyses proved that VIOLET mechanism could ensure non-negative utilities for all users, and resist untruthful behaviors from SUs. Simulation results indicated that VIOLET mechanism can enhance the spectrum allocation efficiency in terms of spectrum utilization, utility of the auctioneer and buyers' satisfaction ratio.

References

1. C. Yi, J. Cai, G. Zhang, Online spectrum auction in cognitive radio networks with uncertain activities of primary users, in *Proceedings of IEEE ICC*, June 2015, pp. 7601–7606
2. A. Subramanian, H. Gupta, Fast spectrum allocation in coordinated dynamic spectrum access based cellular networks, in *Proceedings of IEEE DySPAN*, 2007, pp. 320–330
3. D. Niculescu, Interference map for 802.11 networks, in *Proceedings of ACM SIGCOMM*, 2007, pp. 339–350
4. S. Sodagari, A. Attar, S. Bilen, On a truthful mechanism for expiring spectrum sharing in cognitive radio networks. IEEE J. Sel. Areas Commun. **29**(4), 856–865 (2011)
5. L. Deek, X. Zhou et al., To preempt or not: tackling bid and time-based cheating in online spectrum auctions, in *Proceedings of IEEE INFOCOM*, April 2011, pp. 2219–2227
6. R.P. McAfee, A dominant strategy double auction. J. Econ. Theory **56**(2), 434–450 (1992)
7. X. Zhou, H. Zheng, TRUST: a general framework for truthful double spectrum auctions, in *Proceedings of IEEE INFOCOM*, April 2009, pp. 999–1007
8. M.T. Hajiaghayi, R.D. Kleinberg et al., Online auctions with re-usable goods, in *Proceedings of ACM EC*, 2005, pp. 165–174.
9. J. Bredin, D.C. Parkes, Models for truthful online double auctions, 2012. arXiv preprint arXiv:1207.1360

Chapter 6
Conclusion and Future Research Directions

6.1 Concluding Remarks

Radio spectrum scarcity has become a critical limitation for the development of new wireless equipments, applications, and services. To alleviate such burden, more and more researches have been conducted for improving the spectrum utilization efficiency. In this brief, we focused on one of the most promising technologies, i.e. CR-based dynamic spectrum sharing, from the view of engineering economics. In Chap. 1, the architecture of CR networks and the characteristics of traditional DSA were first presented. Then, the framework of market-driven spectrum sharing was illustrated. As mathematical backgrounds, Chap. 2 reviewed the fundamentals of mechanism design theory, and described some well-known existing mechanisms, such as SPSB, VCG and LOS. After that, three featured spectrum sharing mechanisms were demonstrated in detail. Specifically, a recall-based spectrum auction mechanism was studied in Chap. 3, where a single-seller spectrum sharing model with dynamic spectrum availabilities was considered. In Chap. 4, a two-stage spectrum sharing framework was modeled, in which a multi-seller recall-based spectrum sharing problem was analyzed by a first-stage combinatorial auction mechanism along with a second-stage Stackelberg pricing game. Chapter 5 introduced an online spectrum allocation mechanism, which aims to the scenarios with both PUs and SUs declaring their spectrum usage requests on the fly. In addition, theoretical analyses and numerical results were provided in all these chapters to prove the feasibility, efficiency, and superiority of all aforementioned designs.

© The Author(s) 2016   99
C. Yi, J. Cai, *Market-Driven Spectrum Sharing in Cognitive Radio*,
SpringerBriefs in Electrical and Computer Engineering,
DOI 10.1007/978-3-319-29691-3_6

6.2   Future Works

Market-driven dynamic spectrum sharing with mechanism design approach has been considered as a prospective paradigm in the future realization of CR networks. However, there is still need for designing better spectrum sharing mechanisms which can efficiently increase the spectrum utilization while maintaining desired economic properties. At the end of this brief, we highlight some future research directions in this area.

- *Spatial and temporal spectrum reuse*: Unlike the conventional commodities, radio spectrum can be possibly accessed by multiple wireless users if their mutual interference is negligible. Apparently, enabling such reusability can further increase the spectrum utilization efficiency, and a natural way is to consider the heterogeneities of spectrum requests in spatial and temporal domains.
- *Privacy guarantee in spectrum sharing mechanisms*: In most of the existing spectrum sharing mechanisms, all participants are required to truthfully report their private information, such as values, radio coverages, and geographic locations. However, these information may be sensitive so that participants are reluctant to share them with the auctioneer. Thus, privacy preserving may also be an important desired property for mechanism design in future works.
- *Combination of mechanism design and other methods*: Spectrum sharing problems are generally complicated due to the specific features of radio environments. In the hope of achieving better allocation outcomes or more flexible and effective sharing schemes, it is necessary for us to exploit the advantages of mechanism design and other methods such as by integrating them in a single design.

Printed in the United States
By Bookmasters